Geochemical Controls on Chromium Occurrence, Speciation, and Treatability

Awwa Research Foundation

About the Awwa Research Foundation

The Awwa Research Foundation (AwwaRF) is a member-supported, international, nonprofit organization that sponsors research to enable water utilities, public health agencies, and other professionals to provide safe and affordable drinking water to consumers.

The Foundation's mission is to advance the science of water to improve the quality of life. To achieve this mission, the Foundation sponsors studies on all aspects of drinking water, including supply and resources, treatment, monitoring and analysis, distribution, management, and health effects. Funding for research is provided primarily by subscription payments from approximately 1,000 utilities, consulting firms, and manufacturers in North America and abroad. Additional funding comes from collaborative partnerships with other national and international organizations, allowing for resources to be leveraged, expertise to be shared, and broad-based knowledge to be developed and disseminated. Government funding serves as a third source of research dollars.

From its headquarters in Denver, Colorado, the Foundation's staff directs and supports the efforts of more than 800 volunteers who serve on the board of trustees and various committees. These volunteers represent many facets of the water industry, and contribute their expertise to select and monitor research studies that benefit the entire drinking water community.

The results of research are disseminated through a number of channels, including reports, the Web site, conferences, and periodicals.

For subscribers, the Foundation serves as a cooperative program in which water suppliers unite to pool their resources. By applying Foundation research findings, these water suppliers can save substantial costs and stay on the leading edge of drinking water science and technology. Since its inception, AwwaRF has supplied the water community with more than $300 million in applied research.

More information about the Foundation and how to become a subscriber is available on the Web at **www.awwarf.org**.

Geochemical Controls on Chromium Occurrence, Speciation, and Treatability

Prepared by:
Janet G. Hering
California Institute of Technology, Pasadena CA 91125
and
Thomas C. Harmon
University of California, Merced, CA 95344

Jointly sponsored by:
Awwa Research Foundation
6666 West Quincy Avenue, Denver, CO 80235-3098

Foothill Municipal Water District

Calleguas Municipal Water District

Eastern Municipal Water District

Inland Empire Utilities Agency

Las Virgenes Water District

West Basin Municipal Water District

and

Metropolitan Water District of Southern California

Published by:

Awwa Research Foundation
Advancing the Science of Water®

IWA Publishing

DISCLAIMER

This study was funded by the Awwa Research Foundation (AwwaRF). AwwaRF assumes no responsibility for the content of the research study reported in this publication or for the opinions or statements of fact expressed in the report. The mention of trade names for commercial products does not represent or imply the approval or endorsement of AwwaRF. This report is presented solely for informational purposes.

Published by IWA Publishing, Alliance House, 12 Caxton Street, London SW1H 0QS, UK

Telephone: +44 (0) 20 7654 5500; Fax: +44 (0) 20 7654 5555; Email: publications@iwap.co.uk;
Web: **www.iwapublishing.com**

AwwaRF report number 91043F. Originally published by AwwaRF for its subscribers in 2004.
IWA Publishing version published 2005.

ISBN 1 84339 925 3

CONTENTS

TABLES

FIGURES

FOREWORD

The Awwa Research Foundation is a nonprofit corporation that is dedicated to the implementation of a research effort to help utilities respond to regulatory requirements and traditional high-priority concerns of the industry. The research agenda is developed through a process of consultation with subscribers and drinking water working professionals. Under the umbrella of a Strategic Research Plan, the Research Advisory Council prioritizes the suggested projects based upon current and future needs, applicability, and past work; the recommendations are forwarded to the Board of Trustees for final selection. The foundation also sponsors research projects through the unsolicited proposal process; the Collaborative Research, Research Applications, and Tailored Collaboration programs; and various joint research efforts with organizations such as the U.S. Environmental Protection Agency, the U.S. Bureau of Reclamation, and the Association of California Water Agencies.

This publication is a result of one of these sponsored studies, and it is hoped that its findings will be applied in communities throughout the world. The following report serves not only as a means of communicating the results of the water industry's centralized research program, but also as a tool to enlist the further support of the nonmember utilities and individuals.

Projects are managed closely from their inception to the final report by the foundation's staff and large cadre of volunteers who willingly contribute their time and expertise. The foundation serves a planning and management function and awards contracts to other institutions such as water utilities, universities, and engineering firms. The funding for this research effort comes primarily from the Subscription Program, through which water utilities subscribe to the research program and make an annual payment proportionate to the volume of water they deliver and consultants and manufacturers subscribe based on their annual billings. The program offers a cost-effective and fair method for funding research in the public interest.

A broad spectrum of water supply issues is addressed by the foundation's research agenda: resources, treatment and operations, distribution and storage, water quality and analysis, toxicology, economics, and management. The ultimate purpose of the coordinated effort is to assist water suppliers to provide the highest possible quality of water economically and reliably. The true benefits are realized when the results are implemented at the utility level. The foundation's trustees are pleased to offer this publication as a contribution toward that end.

Walter J. Bishop
Chair, Board of Trustees
Awwa Research Foundation

James F. Manwaring, P.E.
Executive Director
Awwa Research Foundation

ACKNOWLEDGMENTS

The authors of this report are indebted to the following water utilities and individuals for the cooperation and participation in this project:

Calleguas Municipal Water District, Thousand Oaks, CA, Donald Kendall
Inland Empire Utilities Agency, Fontana, CA, Richard Atwater
Las Virgenes Water District, Calabasas, CA, James Colbaugh
West Basin Municipal Water District, Carson, CA, Darryl Miller
Foothill Municipal Water District, La Cañada Flintridge, CA, Ronald Palmer
Metropolitan Water District of Southern California, Los Angeles, CA, Ronald Gastelum
Cadiz, Inc., Santa Monica, Mark Liggett

Laboratory experiments and field sampling associated with this project were conducted by Dr. Giehyeon Lee, Larry Stewart, and Andrew Gadkte (Caltech) and simulations were performed by Mr. Jason Fisher and Ms. Maria Echarte (UCLA). Dr. Lee, Mr. Fisher, and Ms. Echarte also participated in interpretation and analysis of their results and in the preparation of this report. Advice from the AwwaRF project officer, Jason Allen, members of the Project Advisory Committee (PAC), Dr. Philip Brandhuber (HDR), Dr. David Kimbrough (Castaic Lake Water Agency), and Dr. Julie Sueker (Blasland, Bouck & Lee), and from the initial AwwaRF project officer, Dr. Djanette Khiari, are appreciated.

EXECUTIVE SUMMARY

INTRODUCTION

Chromium(VI) is widely recognized as an industrial pollutant and its occurrence in groundwater is often attributed to contamination associated with industries such as chrome plating, metal finishing and aerospace. However, Cr(VI) in groundwater can also derive from natural sources, specifically the weathering of Cr-containing aquifer minerals. The release of Cr from geologic materials can lead to accumulation of Cr in infiltrating water under conditions of either natural or artificial recharge.

The potential accumulation of hazardous solutes, such as Cr(VI), during groundwater recharge and storage may, in part, determine the feasibility of aquifer storage and recovery (ASR) projects. In ASR projects, surplus water is stored below ground for use during periods of excess demand. Arid and semi-arid regions, which are prone to drought and experience wide seasonal variations in water availability and where storage in surface reservoirs is subject to evaporative losses, could greatly benefit from seasonal or long-term groundwater storage. However, changes in water quality during storage have been observed in some ASR projects.

Surveys of the groundwater quality in California, conducted by the State Department of Health Services, have indicated that Cr(VI) concentrations in the tens of μg/L may be quite common even in the absence of any identifiable source of industrial contamination. These values, although below the State and Federal drinking water standards of 50 and 100 μg/L (for total Cr), are higher than those typically found in surface waters that provide the bulk of the municipal water supply in California. Concern over the possible degradation of water quality as compared to the quality of the historical water supply rather than to State or Federal drinking water standards has limited the use of some groundwater supplies in California.

The goal of this project is to improve the science base for groundwater resource management, particularly for groundwater supplies subject to naturally-elevated concentrations of Cr(VI). The project focuses on Cr geochemistry and its influence on the occurrence, speciation, and treatability of Cr in groundwater. Laboratory experiments and numerical simulations were performed to assess the mechanisms and rates of processes by which Cr is partitioned between dissolved and solid phases (in both natural and treatment systems) and to predict the potential for Cr accumulation in recovered water in ASR projects.

RESEARCH OBJECTIVES AND APPROACH

This project addressed a proposed conjunctive use site where naturally-occurring Cr(VI) has been reported to exemplify the application of reactive transport modeling to the assessment of ASR feasibility. Since the exact source of Cr(VI) at this site is not known, simulations were used to examine the effects of sourcing Cr from the saturated or unsaturated zone and of the rates of Cr release from geologic materials on the predicted Cr concentrations in recovered water (i.e., water pumped from ASR extraction wells). The sensitivity of recovered water quality (and hence ASR efficiency) to parameters such as dispersivity, recharge rate, and recharge configuration were tested with simulations that used a single computational domain to model the unsaturated and saturated zones simultaneously. The prospect of reducing the computational burden by decoupling the unsaturated and saturated zone simulations was examined using a dual-domain approach. This approach required the development of an innovative numerical scheme to match the grids at the unsaturated-saturated interface such that the effluent flux from the small-scale, high resolution finite element grid used for the unsaturated zone could be fed to the large-scale, low resolution grid used for the saturated zone. In general, simulations relied on literature values for the weathering of silicate minerals, considered as the host phase for Cr; this process was further examined in laboratory studies.

Experimental studies were conducted with several objectives. The rates of dissolution and Cr-release were examined for fuchsite, a Cr-containing aluminosilicate mineral, as a model for host silicate phases, amorphous Cr(III) hydroxide as a model for Cr(III) coatings on aquifer minerals, and an aquifer sediment sample obtained from the proposed conjunctive use site. Since Cr occurs in minerals primarily in the +III oxidation state, Cr release was examined under both non-oxidizing and oxidizing conditions; for convenience in the laboratory, oxidizing conditions were maintained by addition of the strong oxidant hypochlorous acid/hypochlorite ($HOCl/OCl^-$) in excess. Under oxidizing conditions, oxidation of Cr(III) to Cr(VI) would be expected to occur concurrently with the release of Cr into solution. Under non-oxidizing conditions, Cr would be released from the solid as Cr(III) and might be subsequently oxidized (e.g., by reaction with manganese oxides in the aquifer sediments). Sorption of Cr(VI) to fuchsite and amorphous Cr(III) hydroxide was also examined since this process could influence solid dissolution under acidic conditions. A geochemical model was used to examine the processes of dissolution-precipitation and sorption-desorption in greater detail than was incorporated into the reactive transport simulations. Additional experimental studies were performed to assess the feasibility of redox-assisted coagulation for removal of Cr(VI) from potable water and to examine Cr speciation.

PRINCIPAL FINDINGS OF EXPERIMENTAL STUDIES

Redox-assisted coagulation was found to be an effective method to lower Cr(VI) concentrations in natural waters from 50 to <0.5 μg/L. Removal efficiencies were affected by pH and water composition but could be improved by increasing the Fe(II) dose. Reduction of Cr(VI) to Cr(III) is a critical step in Cr(VI) removal by Fe(II) salts. Cr(VI) sorption onto Fe(III) oxyhydroxides plays a negligible role in Cr removal, particularly above pH 7. Cr(VI) removal by Fe(II) salts was most efficient at pH 7-8. At lower pH values, Cr(VI) removal (over a fixed time interval) was limited by the kinetics of the reaction of Cr(VI) with Fe(II), which decreases with decreasing pH in this range. At higher pH, Cr(VI) removal is limited by the loss of Fe(II) by its competing reaction with dissolved oxygen.

Filterability experiments with colloidal suspensions of $Cr(OH)_3(s)$ demonstrated that a significant fraction of Cr passing through 0.2 and 0.02 μm filters was colloidal and could be retained using an ultrafilter with a nominal pore size of 0.003 μm. The Cr concentration measured in the filtrate obtained using the ultrafilter was in reasonable agreement with the predicted concentration of dissolved, inorganic Cr(III) based on equilibrium with the solid. In field samples, the consistency of Cr measured in unfiltered, 0.2-μm filtered, and ultrafiltered samples indicated that the contribution of a colloidal fraction was negligible in these samples.

In weathering experiments with fuchsite, release of the major constituents Al and Si and the trace constituent Cr were most congruent at pH 9 under oxidizing conditions and least congruent at pH 9 under non-oxidizing conditions. Steady-state release rates (normalized to the elemental composition of the solid) were slower for Cr than for Al and Si under non-oxidizing conditions at both pH 9 and 2. Under non-oxidizing conditions, release of Al and Si were not entirely congruent. Preferential release of Al was observed at pH 2 and preferential release of Si at pH 9. However, rates of release of these major constituents were generally consistent with values reported in the literature for muscovite dissolution. Under oxidizing conditions, Si and Al were released somewhat more slowly than under non-oxidizing conditions. It is possible that the oxidation of Fe(II) in the silicate matrix contributed to this effect but this hypothesis remains to be tested.

In dissolution experiments with $Cr(OH)_3(s)$, dissolved Cr concentrations were observed to increase linearly with time in batch experiments conducted at pH 9. The accumulation of Cr in solution corresponded to dissolution of 40% of the solid in 200 min. Release of Cr was negligible under non-oxidizing conditions at this pH. At pH 2, some dissolution of the solid was observed under oxidizing conditions (though much less than at pH 9). The rate of the reaction at pH 2 decreased with time, however, and the dissolved Cr concentration appeared to reach a plateau. Under non-oxidizing conditions, Cr release was more nearly linear but also slower than under oxidizing conditions. Sorption of Cr(VI) onto $Cr(OH)_3(s)$ is strongly pH dependent with negligible sorption at pH 9 and increasing sorption with decreasing pH. The sorption isotherm obtained at pH 3 suggests that, under the conditions of the dissolution experiment at pH 2 under oxidizing conditions, the $Cr(OH)_3(s)$ surface would rapidly become saturated with Cr(VI). This may inhibit the further oxidative dissolution of the solid.

Release of Al, Si, and Cr from aquifer sediments was examined in batch experiments only at pH 9 but under both oxidizing and non-oxidizing conditions. The Cr content of the aquifer sediment was at the detection limit for the XRF analysis (0.01%). Dissolution rates could not be quantified under non-oxidizing conditions. Under oxidizing conditions, release of Cr from the solid was observed despite its low Cr content. Cr concentrations increased linearly with time but reached only 4 μg/L in 14 days.

Geochemical modeling was performed to examine the evolution of water composition with time in two cases. In the first case, dissolution of a primary solid was considered alone or with precipitation-dissolution of a secondary solid and/or sorption-desorption of the dissolved constituent. Dissolution of the primary solid results in the linear increase in the dissolved concentration of its constituent elements (here called X). Precipitation-dissolution of a secondary solid (also containing X) results in the dissolved concentration of X reaching a steady-state value (even as the dissolution of the primary solid continues). Sorption-desorption of X (under conditions such that the concentration of surface sites are limited) slows the accumulation of X in the dissolved phase but does not change the eventual steady-state concentration of X (which is determined by pseudo-equilibrium with the secondary solid). In the second case, oxidative dissolution of $Cr(OH)_3(s)$ was considered with subsequent adsorption of the product Cr(VI); sorbed Cr(VI) was also allowed to inhibit the further oxidative dissolution of the solid. In the oxidative dissolution of $Cr(OH)_3(s)$, rapid saturation of surface sites is predicted (i.e., for pH conditions favoring Cr(VI) adsorption). Inclusion of an inhibitory effect of sorbed Cr(VI) provides a reasonable model for the observed release of Cr from $Cr(OH)_3(s)$ under oxidizing conditions at pH 2. Sorption alone (i.e., without inhibition) has little effect on the accumulation of Cr in solution once the surface is saturated.

PRINCIPAL FINDINGS OF NUMERICAL STUDIES

Soil and groundwater flow and transport simulations using the numerical models FEMWATER and HGBC123D were used to evaluate the potential field-scale effects of natural chromium sources on ASR operations. Two sets of model simulations were undertaken. The first and more extensive simulations employed a single computational domain to model the unsaturated and saturated zones simultaneously. A second set of simulations was undertaken to test the prospect of decoupling the unsaturated and saturated zone simulations to provide a more detailed account of the unsaturated zone while alleviating computational burden. In all simulations, ASR water quality was estimated for local (recovery well) behavior, and global behavior as summarized by spatial moments estimates of the simulated spatial chromium concentration distributions.

Single Domain Simulations

In the single-domain simulations, three scenarios were tested: (I) Preexisting background Cr concentrations in the aquifer (before ASR operations begin) but no dissolution/weathering source, (II) Preexisting Cr concentrations and zero-order dissolution/weathering sources uniformly distributed in the aquifer, and (III) Zero-order dissolution/weathering in the unsaturated zone only. For Scenario I, key parameters were identified and selected for testing the sensitivity ASR efficiency in terms of recovered water quality. These parameters included dispersivity, recharge rate and recharge configuration. Scenario II focused on the impact of the weathering rate on recovered water quality, invoking the range of zero-order rate constants gleaned from the literature on silicate weathering (Chapter 1). Scenario III again focused on the weathering rate constants, but from the perspective of the potential impact of unsaturated zone weathering products on recovered water quality.

The advantage of the single-domain approach is that it captures the true dynamics of the unsaturated and saturated flow processes. More specifically, the single-domain approach captures the mounding dynamics which occur when the recharging water reaches the water table. The mounding in turn dominates local vertical and horizontal flow patterns in the aquifer.

Parameter sensitivity analyses indicated that the greatest changes in the simulated amount of clean water recovered (i.e., ASR efficiency) resulted from changes in recharge rate and recharge configuration. This result is a function of the mounding effect on mixing between the clean (recharged) and native groundwater. The horizontal movement and trajectory of the clean water plume was most affected by recharge rate and duration of the recharge period. Concentration time series at observation wells downgradient from the recharge zone were most sensitive to the value of dispersivity.

For the domain and conditions simulated, the impact of dissolution rate on normalized local concentrations downgradient are affected by the value of incoming concentration. Incoming concentration had a significant effect at the low reaction rates, but the effect lessened with increasing reaction rate. There was an upper bound for the normalized concentration when dissolution into the recharged water took place against a background that had reached steady state. Cr weathering/dissolution rates of 1e-12 mol/kg/d and lower were sufficiently slow to produce output that closely resembled that for the non-reactive transport base. This means that the dissolution kinetics could be negligible over the time scale associated with these cases. For reaction rates of 1e-9 mol/kg/d and greater, local equilibrium was valid throughout the simulated domain. Intermediate rates (1e-12 mol/kg/d < k < 1e-9 mol/kg/d) produced significant effects on the recovered water quality for the conditions simulated. The results from the literature suggest that silicate weathering rates are extremely low, ranging from 10^{-17} to 10^{-11} mol/m^2 s. Given that the stoichiometric content of chromium in most silicate minerals is expected to relatively low (< 1%), congruent chromium dissolution rates may be two or more orders of magnitude lower than this range, suggesting that Cr weathering would not impact ASR water significantly for the geometry and timescales studied (roughly 10-15 year storage periods). However, mixing with the native water will always impact the ASR water quality as noted above.

The application of spatial moment analysis was demonstrated in this work as a robust method for quantifying clean plume behavior in the context of an ASR setting. Spatial moments provide an integrated assessment of simulated ASR efficiency or clean-water recovery (zeroth moment), clean-water plume centroid location (first moments with respect to different axes), and clean-water plume travel time (first moment derivative with time).

The disadvantage of the single-domain approach (and motivation for the dual-domain approach) is that accuracy of the simulation is highly suspect in the absence of a high-resolution spatial and temporal discretization in the unsaturated portion of the domain. The problem with this is that the overall scale of the combined saturated and unsaturated grids leads to excessive computation burden. The greater computational requirements are attributed to the highly nonlinear form the governing equations take under unsaturated conditions. These nonlinearities are dealt with by increasing both the spatial and temporal resolutions, thus increasing the computational time required for a simulation. In the simulations used here, the "fine" resolution simulations were actually coarse in an absolute sense in order to allow simulation times ranging from 30 min to a few hours. However, trial simulations focusing on detail in the unsaturated zone drove these simulation times up to tens of hours or even days on PC Pentium 4 processors operating at 2.5 GHz.

Dual Domain Simulations

The dual-domain simulations required numerical innovations and were undertaken using the same parameters as in the base simulations for the single-domain approach. Innovations required were associated with using a small-scale, high-resolution finite element grid for the unsaturated domain, then feeding the effluent flux from this domain to a large-scale, low-resolution grid in the saturated domain. The resulting numerical scheme for matching the grids at the unsaturated-saturated interface is referred to as the recharge boundary construction algorithm.

The dual-domain simulations based on the boundary recharge construction algorithm were successful if the value of the key numerical error criterion (maximum % difference) selected was sufficiently small. A tradeoff was identified between this criterion and the computational times required for the dual-domain simulations. For the domains and conditions of this study, computation times ranged from greater than 10 to less than 3 hours with decreasing levels of stringency.

RECOMMENDATIONS TO UTILITIES

Groundwater storage offers substantial benefits to utilities that seek additional flexibility in managing their water supply. However the use of this water management strategy may be compromised if the quality of recovered water is degraded relative to that of the water used for recharge. In some cases, groundwater storage will only be feasible if the recovered water can be distributed to consumers with only minimal treatment (e.g., disinfection). Even when recovered water is delivered to a treatment facility prior to distribution, some concerns may arise if water quality is degraded during storage in the subsurface.

If treatment is required to lower Cr(VI) concentrations, several technologies, including ion exchange and reverse osmosis, can be applied. Here, bench-scale tests demonstrated that sub-ppb levels of Cr(VI) can be achieved by addition of Fe(II) salts even in the presence of dissolved oxygen. The mechanistic basis for this redox-assisted coagulation process is the reaction of Cr(VI) with Fe(II) to produce Cr(III) and Fe(III). Hydrolysis and precipitation of the products results in incorporation of Cr(III) in a mixed Cr(III)-Fe(III) hydroxide solid, which can be removed by filtration. Treatment efficiency can be improved by pH optimization and by increasing the Fe(II) dose. These tests were run on waters spiked with Cr(VI) and would need to be validated for source waters containing elevated Cr concentrations. Although tests were not run in the absence of dissolved oxygen, previous studies demonstrate that the process would also be effective under those conditions.

For water samples in which the total Cr concentrations in conventionally filtered samples exceed the measured Cr(VI) concentrations, assessment of the possible contribution of colloidal Cr would be particularly important. The presence of colloidal Cr(III) was detected in laboratory studies by comparing the total Cr concentrations measured in ultrafiltered samples with those measured in filtrates passed through 0.2 μm filters. A similar test run with samples collected in the field demonstrated that colloidal Cr was not present in these field samples.

The rate of Cr release from aquifer minerals will be influenced by whether the oxidation of Cr(III) to Cr(VI) occurs concurrent with or subsequent to mineral dissolution. The extent of this effect will depend on the type of mineral hosting Cr in the soil or aquifer sediment. Cr release from silicate minerals is very slow and is congruent with the release of the major elements Al and Si only under oxidizing conditions. Under non-oxidizing conditions, Cr release is even slower than the dissolution of the mineral framework. If Cr is present as hydroxide coatings on aquifer minerals, however, its release will be faster than if is hosted in aluminosilicate minerals. Release of Cr from Cr(III)hydroxide solids is strongly enhanced under oxidizing conditions at basic pH. At acidic pH, however, the product Cr(VI) adsorbs onto the residual solid, which appears to inhibit further dissolution.

Simple geochemical modeling can be used to illustrate the effects of dissolution-precipitation and sorption-desorption on the accumulation of dissolved constituents over time under batch reactor conditions. In particular, sorption can potentially influence the rate of accumulation of a dissolved constituent, either through its loss from solution to sorption sites or the inhibition of the reaction of the underlying solid. However, if the concentration of sorption sites is limited, the steady-state concentration of dissolved constituents will not ultimately be limited by sorption.

Any proposed ASR project should be accompanied by a modeling investigation aimed at predicting flow and transport effects under anticipated operating conditions. The general procedures for model construction, discretization and parameter sensitivity analysis are described in this work, but other ASR projects will undoubtedly be subject to different scales and conditions and may have different operational horizons.

Single-domain ASR simulations are a viable modeling alternative if resolution of results in the unsaturated zone is not important. This approach is preferable, especially if the simulation model is to be coupled to a water resources management algorithm requiring repetitive simulations. Cases clearly exist where better unsaturated zone resolution is needed. For example, proximate groundwater-surface water interactions, such as at streams or springs, may be impacted by the recharge and extraction rates, and by input/output timing of the ASR operation.

If resolution in the unsaturated zone is a requirement, or if the model is to also serve as a water resources management tool (placing the model within the framework of an optimization algorithm), then the additional computational costs associated with a high resolution single-domain model would become unacceptable. In this case it is better to explore the use of the dual-domain approach described in this study.

The potential impact of the natural chromium release rates likely to govern in the case of silicate weathering processes were shown here to be minimal for the conditions and timescales simulated here. This means that release rates identified in the literature were sufficiently slow relative to the clean-water residence time in this system. This result means that conservative transport simulation using FEMWATER would suffice and the more complex HBGC123D model could be avoided. It is important to note that this result is linked to the scale of the ASR system and the ASR operational parameters governing the clean-water residence time. An ASR project with a greater residence time, faster weathering rates, or some combination of the two, may demonstrate significant chromium introduction into clean-water plumes during the storage cycle of an ASR operation.

CHAPTER 1
BACKGROUND

CHROMIUM REGULATORY STANDARDS AND HEALTH EFFECTS

The maximum contaminant levels (MCLs) for total Cr in drinking water are 100 µg/L for the U.S. and 50 µg/L for California. At the current time, Cr(VI) is not regulated separately from total Cr but California state law requires that an MCL specific to Cr(VI) be set in 2004 (DHS, 2003). This requirement reflects the fact that Cr(III) is an essential nutrient with a known role in glucose metabolism (Vincent, 2000). An organically-complexed form of Cr(III), chromium picolinate, is a popular dietary supplement and has been used (as an insulin sensitizer) in the treatment of diabetes (Anderson, 2000). Hence the fraction of total Cr in drinking water that is present as Cr(III) is not considered to pose any hazard to human health.

Cr(VI) is a known human carcinogen by inhalation of sparingly soluble compounds such as calcium, lead, strontium, and zinc chromate salts and chromium trioxide (CrO_3). There is, however, considerable debate regarding the carcinogenicity of Cr(VI) by ingestion (Flegal et al., 2001; Kimbrough et al., 1999). Animal studies have generally failed to show carcinogenic effects of ingested Cr(VI). The Federal MCL is based on the accumulation of Cr in animal tissues following ingestion of Cr(VI) as a toxicological endpoint; the MCL for total Cr is deemed to be protective for Cr(VI) (DHS, 2003). Furthermore, the known capacity of the gut to reduce Cr(VI) to Cr(III) is expected to have a protective effect when Cr(VI) is ingested. A recent study with human volunteers in which Cr uptake by red blood cells was used as the metric for Cr(VI) adsorption demonstrated that the reductive capacity of the stomach is not overwhelmed by ingestion of Cr(VI) in drinking water at mg/L levels (Paustenbach et al., 2003). The results of an on-going study of the carcinogenicity of ingested Cr(VI) in rodents being conducted by the National Toxicity Program may help to resolve the question of Cr(VI) carcinogenicity (NTP, 2003).

CHROMIUM OCCURENCE IN GROUNDWATER

In California, concern over the occurrence of chromium(VI), Cr(VI), in groundwater has been based primarily on contamination from industrial sources. Land disposal of "blow-down" water from cooling towers containing Cr(VI) as a corrosion inhibitor caused significant groundwater contamination in Hinkley, CA (ATSDR, 2003). Numerous sites with chromium contamination have been identified in the Los Angeles Basin; these are associated with industries such as chrome plating, metal finishing and aerospace, and include the San Fernando Valley Superfund site contaminated by releases from a former Lockheed facility (Blankstein and Guccione, 2000; Guccione, 2000). Cr(VI) has been found in groundwater at concentrations up to 580 µg/L in Hinkley and up to 408 µg/L in Los Angeles County (Martin and Blankstein, 2001).

Cr(VI) in groundwater can, however, also derive from natural sources. The average crustal abundance of chromium is 100 ppm (Klein and Hurlbut, 1993) and significant enrichments over crustal abundance occur in mafic and ultramafic igneous and meta-igneous rocks (Fantoni et al., 2002). Chromium occurs in geologic materials, however, primarily in the +III oxidation state thus, if these materials are the source of Cr(VI) in groundwater, oxidation of chromium from Cr(III) to Cr(VI) must occur either concurrent with or subsequent to the release of chromium from the solid phase.

There are several examples, mostly from arid or semi-arid regions, of natural occurrences of elevated Cr(VI) in groundwater. In Italy, Cr(VI) concentrations ranged from 5 to 73 μg/L with higher concentrations found in $Mg-HCO_3$ type waters than in $Ca-HCO_3$ type waters. This pattern was attributed to weathering of different minerals, periodites (with 1800 ppm Cr) for Mg-rich waters and basalts (with 185 ppm Cr) for Ca-rich waters (Fantoni et al., 2002). In Argentina, total Cr concentrations ranging from 20-230 μg/L were found in aquifers dominated by aeolic deposits with substantial contribution of volcanic materials (Farias et al., 2003). In Arizona, Cr(VI) was found in groundwater at concentrations up to 200 μg/L. The highest concentrations were associated with fine grained sediments and elevated pH (up to ca. 9). However, total Cr concentrations in the clay-silt fraction were not elevated (mean of 65 ppm) (Robertson, 1975). In Mexico, ultramafic sources of Cr were identified in geologic materials. Discrete chromite contained up to 4% Cr and serpentinites up to 4000 ppm Cr. However, total Cr concentrations in groundwater were below 15 μg/L and the fraction of total Cr as Cr(VI) was highly variable (Robles-Camacho and Armienta, 2000).

In testing conducted by the California Dept. of Health Services, Cr(VI) has been detected in groundwater at concentrations ranging from non-detectable (<0.5 μg/L) to 35 μg/L and constituting between 8 and 100% of total Cr (Table 1.1). Further sampling of over 6400 water sources found >10 μg/L Cr(VI) in over 600 sources and >50 μg/L Cr(VI) in 5 sources. Where Cr does occur primarily as Cr(VI), the oxidation of Cr(III) to Cr(VI) has been attributed to reaction with dissolved oxygen or manganese(III,IV) oxides (Fantoni et al., 2002; Robertson, 1975) but no direct evidence supports these contentions.

Table 1.1
Recent data on Cr occurrence and speciation in California groundwater*

sampling period	number of wells	county	total Cr (μg/L)	Cr(VI) (μg/L)	Cr(VI) (%)
1997-1998	10	Merced	18.8-33.5	16.8-33.0	68-100
1998	3	Los Angeles	3.4-10.3	<0.5-5.1	<14-54
1998	1	Contra Costa	18.7	1.5	8
1999	9	Los Angeles	5.3-15	3.6-11	58-100
2000	3	Yolo	31-54	24-35	44-97
2000	8	San Mateo	11-28	7.4-28	64-100

*source: http://www.dhs.cahwnet.gov/ps/ddwem/chemicals/chromium6/samplingresults.htm

ENVIRONMENTAL CHEMISTRY AND GEOCHEMISTRY OF CHROMIUM

Chromium in groundwater occurs in both the +VI and +III oxidation states. Chromium is unusual in that both Cr(VI) and Cr(III) can be thermodynamically stable in the presence of atmospheric oxygen depending on pH. Under oxidizing conditions, Cr(III) is stable (as Cr^{3+}) at pH < 2 and Cr(VI) is stable at higher pH either as bichromate, $HCrO_4^-$, or chromate, CrO_4^{2-} (Figure 1.1). The chemical properties and reactivities of Cr(III) and Cr(VI) are quite distinct and hence the mobility of chromium will depend strongly on its redox speciation. Although the concentrations of redox-active species formally define the redox potential (pe or E_h) of the system, it is often the case that redox potentials calculated from different redox couples do not agree with each other or with the measured redox potentials (Cutter, 1992; Holm and Curtiss,

1989) unless extraordinary care is exercised (Grenthe et al., 1992). These discrepancies indicate that concentrations of redox-active species reflect the kinetics of operative redox reactions in the system (which establish steady-state concentrations) rather than redox equilibrium. Thus, the kinetics of reaction of chromium species with environmental oxidants and reductants must be considered.

The oxidation of Cr(III) by oxygen is quite sluggish, except in strongly alkaline solution (Rai et al., 1989; Richard and Bourg, 1991) and ref cit). At 20 °C, only 3% of the initial Cr(III) was oxidized by reaction with dissolved oxygen over 30 days and the rate of reaction was insensitive to pH over the range 5.9 to 9.9 (Schroeder and Lee, 1975). In contrast, rapid oxidation of Cr(III) by Mn(III, IV) oxides has been reported (Eary and Rai, 1987; Fendorf et al., 1992b; Fendorf and Zasoski, 1992a; Johnson and Xyla, 1991; Weaver et al., 2002). However, this reaction appears to oxidize only dissolved Cr(III) species and its rate decreases significantly with pH; reaction above pH 4-5 is inhibited by the precipitation of Cr(III) oxyhydroxides at the Mn(III, IV) oxide surface (Fendorf et al., 1992b; Fendorf and Zasoski, 1992a; Weaver et al., 2002). In soils and aquifers, this pathway for Cr(III) oxidation would be limited both by the release of dissolved Cr(III) species from geologic materials and by the precipitation of secondary Cr(III) oxyhydroxide phases. Cr(VI) is thermodynamically unstable under reducing conditions and is rapidly reduced to Cr(III) by reaction with Fe^{2+} or Fe(II)-containing minerals, such as biotite (Buerge and Hug, 1997; Buerge and Hug, 1999; Hug et al., 1997). Thus, reducing groundwaters (which characteristically contain elevated concentrations of reduced iron and manganese) should not contain measurable concentrations of Cr(VI).

Distinctly different mobility of Cr(III) and Cr(VI) in groundwater is expected because of their different solubilities and affinities for mineral surfaces. As Fig. 1.1 illustrates, Cr(III) is quite insoluble even as the amorphous $Cr(OH)_3(s)$. The maximum concentration of dissolved, inorganic Cr(III) in equilibrium with this solid in the neutral pH range is 7.5 μg/L (Ball and Nordstrom, 1998). Even lower dissolved, inorganic Cr(III) concentrations would be predicted if solid solutions of $Cr,Fe(OH)_3(s)$ are considered; such solids solutions have been found to form readily in laboratory systems (Rai et al., 1989; Rai et al., 1987). Cr(III) is also strongly sorbed onto oxide or clay mineral surfaces. The sorption behavior of Cr(III) is typical for cations with sorption increasing with pH (Dzombak and Morel, 1990). Both the sorption and precipitation of Cr(III) could be suppressed by organic complexation but this effect is unlikely to be significant in groundwaters, which are typically low in dissolved organic matter. Thus, mobility of Cr(III) in groundwater (at concentrations exceeding the solubility limit of 7.5 μg/L) is likely to be due to colloidal transport (in which sorbed or precipitated Cr(III) is transported as colloidal particles). Mobility of colloids in groundwater has been documented and shown to influence the transport of particle-reactive chemical species (McCarthy and Zachara, 1989). Colloidal particles may also pass through a 0.45 μm filter and hence be included in the operationally-defined "dissolved" fraction (see below).

In contrast to Cr(III), the concentration of Cr(VI) in groundwater is not expected to be limited by the solubility of solid chromate-containing phases. Although chromate minerals, such as crocoite ($PbCrO_4(s)$) do occur in nature, they are generally found in evaporite deposits (see below) and would not be predicted to precipitate from dilute solution. Cr(VI) does sorb onto oxide minerals with typical anionic behavior (decreasing sorption with increasing pH) but Cr(VI) sorption is strongly suppressed in the presence of competing anions, such as sulfate, especially at circumneutral pH (Zachara et al., 1987). However, reversible sorptive retardation of Cr(VI) has been observed in small-scale tracer tests conducted in a relatively clean aquifer zone (oxic conditions, mildly acidic with a low dissolved salts concentration) (Kent et al., 1995). In a sewage-contaminated zone within the same aquifer (mildly reducing conditions, slightly higher pH and moderate dissolved salts concentration) a weaker retardation of Cr(VI) was reported. In the latter case, this species was partially lost from the aqueous phase due to reduction to Cr(III) and subsequent irreversible sorption/precipitation (Kent et al., 1994). Similar observations have been documented in a larger scale tracer test executed by the same group in a nearby location. In the larger scale test, spatial trends were also observable as the extent of adsorption decreased vertically in response to increasing pH and concentration of groundwater oxyanions, and Cr(VI) reduction took place quickly in the suboxic zone whereas the rate was slower in the oxic region of the sewage contaminated zone (Davis et al., 2000). In an aquifer at nearly neutral pH, plume migration studies revealed sorption of Cr(VI) as well as reduction to Cr(III) even under aerobic conditions (Henderson, 1994).

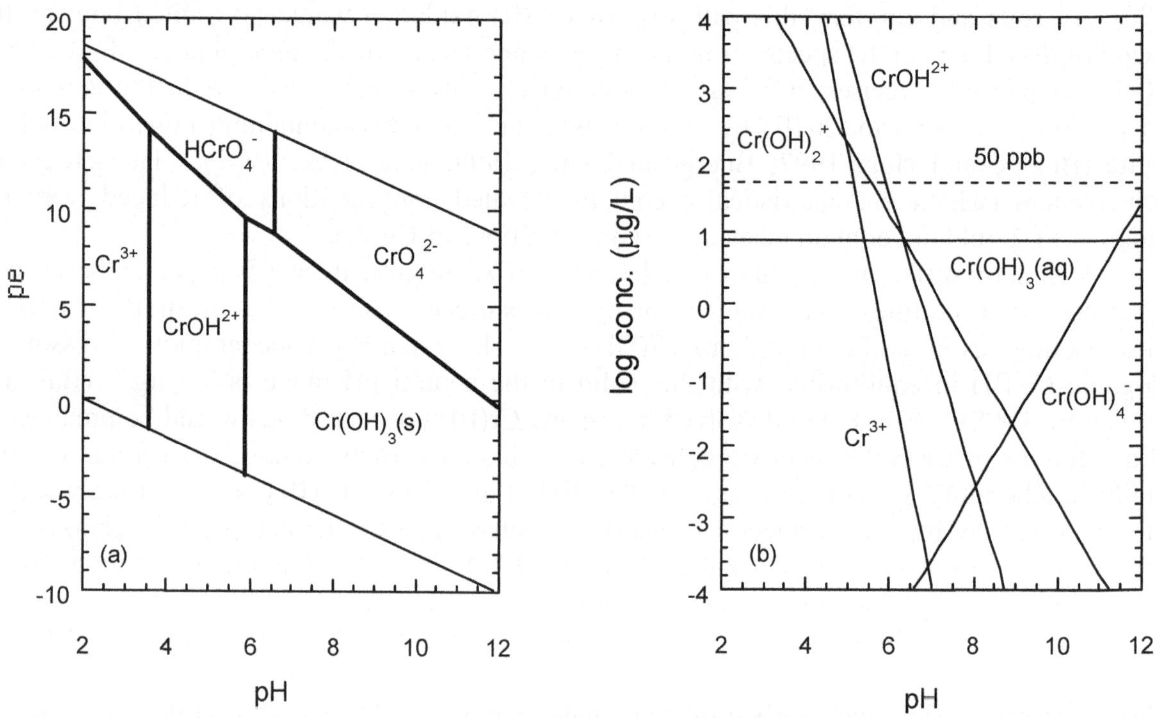

Figure 1.1 Chromium speciation (a) pe-pH diagram for 2 ×10^{-6} M (104 µg/L) total Cr, (b) dissolved Cr(III) species in equilibrium with amorphous $Cr(OH)_3$(s)

ANALYTICAL CHEMISTRY OF CHROMIUM

The EPA-approved protocols (FR, 1995) for analysis of chromium in potable water involves sample filtration followed by either: (1) acidification and analysis for total Cr by AAS, ICP-MS, or other methods within 6 months or (2) adjustment of sample pH to 9 followed by analysis for Cr(VI) by ion chromatography with post-column derivitization and colorimetric detection within 24 hours (with storage at 4 °C) as shown in Figure 1.2. Such determinations provide information needed to assess compliance with the MCL for Cr (which applies to "dissolved" Cr in Figure 1.2) and health risks which are presumed to be associated with exposure to inorganic Cr(VI). In this scheme, the contribution of Cr(III) is determined indirectly by subtracting the concentration of Cr(VI) (quantified by colorimetry) from the "dissolved" Cr concentration and hence would <u>include</u> both colloidal and organic Cr.

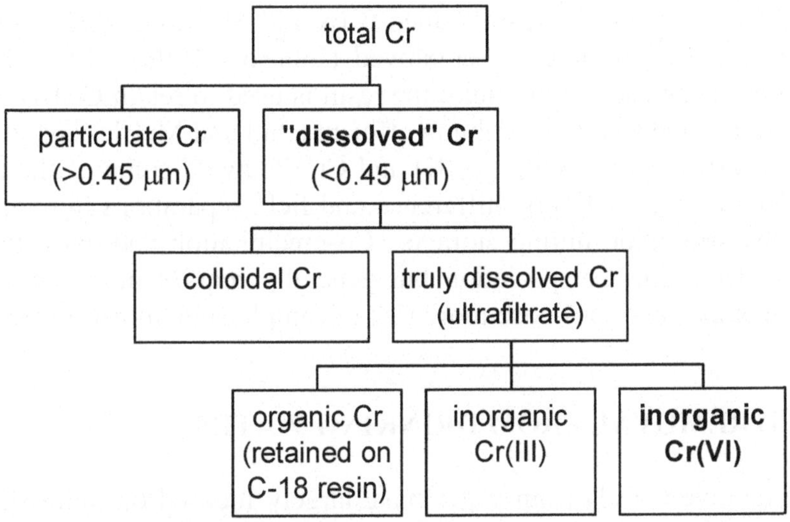

Figure 1.2 Schematic diagram showing distribution of Cr species. Bold text indicates those Cr species determined in standard analyses of potable water.

Organic Cr in natural waters has been determined by collection of Cr onto a hydrophobic resin and was found to be correlated with the dissolved organic carbon (DOC) concentration (Kaczynski and Kieber, 1994). The concentrations of organic Cr concentrations exceeded that of total, inorganic Cr in high-DOC surface freshwaters. Increases in the organic Cr concentrations was observed upon spiking water samples with either Cr(III) or Cr(VI). It was suggested that formation of organic Cr species with Cr(VI) involved its reduction to Cr(III) but the oxidation state of Cr in the organic form was not identified.

More commonly, analytical distinctions are made between Cr(III) and Cr(VI) but all "dissolved" Cr is assumed to be inorganic. Organic Cr is less likely to be important in low-DOC groundwaters than in high-DOC surface waters, thus this assumption may be reasonable for groundwater systems. However, neglect of colloidal Cr species is more problematic, particularly if the difference between reported total, "dissolved" Cr and Cr(VI) exceeds the solubility limit for Cr(III). Contributions of the colloidal size fraction to "dissolved" metal concentrations can be rigorously assessed by ultrafiltration (Benoit and Rozan, 1999; Wen et al., 1996), field-flow filtration (Murphy et al., 1993; Taylor et al., 1992) or photon correlation spectroscopy (Ledin et al., 1995). Such assessment has not been made for chromium in groundwater.

The protocol for sample handling and storage for determination of Cr(VI) also poses some potential problems. Significant inter-conversion of Cr(III) and Cr(VI) have been observed during storage of spiked water and wastewater samples (Pavel et al., 1985). Loss of Cr(VI) was observed within days at low pH but could be prevented by adjusting the pH to 9. However, in samples with an excess of Cr(III) over Cr(VI), the concentration of Cr(VI) was observed to increase over time in samples buffered to pH 9 unless the Cr(III) was masked by addition of the strong organic complexing agent EDTA. The 24-hour holding time also imposes fairly significant practical burdens on utilities subject to monitoring requirements.

Assuming that organic Cr can be neglected, several chromatographic techniques have been used to separate Cr(III) and Cr(VI); some of these methods can be used for field separations. Microcolumns of activated alumina have been used to retain selectively Cr(III) under acidic conditions and Cr(VI) under basic conditions; a good mass balance between total Cr and the sum of Cr(III) and Cr(VI) was obtained (Cox and McLeod, 1992). Anion- and cation-exchange columns have been similarly employed (Johnson, 1990). More recently, a cation exchange method has been used in the field; the resin is used to retain Cr(III) quantitatively and Cr(VI) is measured (as total Cr) in the column effluent (Ball and McCleskey, 2003). Techniques that allow direct measurement of both Cr(III) and Cr(VI) avoid some of the potential artifacts associated with determining Cr(III) by difference and field separations minimize the possibility of changes in redox speciation during storage. Co-precipitation and extraction methods have also been used to determine Cr speciation but generally require more sample handling than chromatographic methods and thus pose more risk of sample contamination (Kotas and Stasicka, 2000).

CHROMIUM(VI) REMOVAL FROM DRINKING WATER

Efforts to remove Cr(VI) from water have largely focused on industrial wastewaters or contaminated groundwaters rather than on potable water treatment. However, some methods for Cr(VI) removal may be useful in a wide range of applications. Methods of Cr(VI) removal generally fall into one of two categories: removal of Cr(VI) as the anion or reduction of Cr(VI) to Cr(III). Cr(VI) can be directly removed from aqueous solutions by anion-exchange resins (Rengaraj et al., 2003; Yalcin et al., 2001) or zeolites (Tahir et al., 1998; Zhang and Reardon, 2003), which have been proposed for recovery of Cr(VI) from electroplating solutions and Cr(VI) removal from tannery waste streams. Since Cr(III) can be easily immobilized by sorption and/or precipitation, reduction of Cr(VI) can be an effective strategy for its removal from water. Cr(VI) can be reduced microbially (Tokunaga et al., 2003; Vainshtein et al., 2003) or by reaction with zero valent iron (Khan and Puls, 2002; Lee et al., 2003; Powell et al., 1995; Puls et al., 1999), organic matter (Kratochvil et al., 1998; Reddad et al., 2003; Wittbrodt and Palmer, 1996), or iron(II) (Eary and Rai, 1988). Several studies have demonstrated that dissolved Cr concentrations in the range of 50 to 100 μg/L could be achieved by treatment of Cr(VI)-contaminated groundwaters with ferrous salts (Aktor, 1994; El-Shoubary et al., 1998; Philipot et al., 1985; Sorg, 1979; Zotter and Licsko, 1992).

WATER QUALITY ISSUES IN AQUIFER STORAGE AND RECOVERY

Aquifer storage and recovery (ASR) is a water management strategy in which surplus water is stored below ground for use during periods of excess demand (Domenico and Schwartz, 1998; Pyne, 1995). In arid and semi-arid regions, ASR has obvious advantages over storage in surface reservoirs, which can be subject to considerable evaporative losses. These regions are also prone to drought and experience wide seasonal variations in water availability (i.e., from precipitation or snowmelt) and could thus benefit from seasonal or long-term storage.

In California, a wide variety of groundwater storage projects (variously referred to as ASR, conjunctive use, water banking, recharge, or storage) have been developed (Table 1.2). These projects serve both municipal and agricultural users and, in some cases, provide control of seawater intrusion. Typical recharge capacities range from tens of thousands to hundreds of thousands of acre-feet annually. Both local runoff and imported water, and less commonly reclaimed water, are used for recharge, which is accomplished largely through spreading basins (Jones, 2003).

Prospects for ASR have also been widely explored and to some extent implemented in the southeastern coastal states, particularly in Florida where ASR is an important component of the Comprehensive Everglades Restoration Plan (Reese, 2002). Unlike in California, many of the aquifers targeted for groundwater storage projects are non-potable aquifers and recovery is often limited in practice by elevated chloride concentrations in the recovered water (Parkhurst and Petkewich, 2002; Reese, 2002).

In addition to the benefit of storage capacity and equalization of water supply, groundwater storage also offers the potential for improvement of water quality. Groundwater recharge has been found to be an effective removal process for pharmaceuticals (Reinhard et al., 2003). Removal of disinfection byproducts and their precursors has also been observed to occur under aquifer storage conditions (McQuarrie and Carlson, 2003). In an agricultural region where shallow groundwater was contaminated with nitrate and atrazine, recharge of surface water resulted in significant decreases in the level of groundwater contamination (Ma and Spalding, 1997).

Concerns over groundwater storage tend to focus on recovery efficiencies, possible changes in hydrologic properties, and subsidence. Increased aquifer permeability during ASR operations has been attributed to dissolution of carbonate minerals and amorphous silica (Mirecki et al., 1998) and reduced permeability to clay dispersion and migration of interstitial clay particles with displacement of brackish formation water by recharged freshwater (Konikow et al., 2001). More recently, however, attention has focused on the potential for degradation of water quality during groundwater storage and some operating ASR projects have been compromised by accumulation of arsenic in the recovered water (Elison, 2003).

Potential for Evolution of Water Quality During ASR

Groundwater storage and, for projects involving surface recharge, infiltration through the vadose zone provide for contact between the recharge water and soil and aquifer minerals that can result in the evolution of the chemical composition of the recharge water. The same processes of water-rock interaction account for the accumulation of solutes and alkalinity by infiltrating water in natural recharge. In the case of artificial recharge, however the recharge water may already carry a significant solute load, which may both influence the extent of and be modified by further water-rock interactions.

Evolution of the chemical composition of recharge water can result from the processes of dissolution-precipitation, ion exchange, and sorption-desorption (Drever, 1997; Langmuir, 1997). For primary minerals (formed at high temperature and pressure), dissolution is effectively irreversible under aquifer conditions. Secondary minerals may both precipitate and dissolve; the extent and rates of these processes will depend on the chemical composition of the recharge water and on the type of soil and aquifer minerals present. Dissolution-precipitation processes can affect the concentrations of both major and trace elements. Ion-exchange processes are generally non-specific and influence largely the major ion composition of the infiltrating water. Trace element composition can be significantly affected by sorption processes though sorption capacity can be saturable, which would limit its effect over time.

Table 1.2
California groundwater storage projects*

agency	uses	designation	capacities	recharge method	comments
Alameda County Water District	municipal seawater intrusion	conjunctive use	avg. recharge: 25,000 AF/yr 16 extraction wells	surface recharge injection	initiated in 1970's for control of seawater intrusion
Arvin-Edison Water Storage District	agricultural municipal	conjunctive use banking	1,500 acres of basins 70 extraction wells banking storage: 350,000 AF	surface recharge	initiated in 1960's for agricultural users
Calleguas Municipal Water District	municipal	ASR	300,000 AF of storage planned 30 wells	injection	initiated in 1990's
City of Bakersfield	municipal agricultural	conjunctive use	1,470 acres of basins est. max. recharge: 168,000 AF/yr	surface recharge	water rights acquired in 1978
Coachella Valley Water District	municipal	storage conjunctive use	typical recharge: 61,000 AF/yr addl recharge by MWDSC: up to 600,000 AF	surface recharge	recharge of imported water initiated in 1973
Kern County Water Agency	agricultural	storage	20,000 acre recharge site, 30 extraction wells	surface recharge	initiated in 1995
Kern Water Bank Authority	agricultural	storage banking	1200 acres of basins, 50 extraction wells est. storage capacity 1 MAF	surface recharge	recharge of local and imported water initiated in 1995
Los Angeles County	municipal seawater intrusion	storage	2400 acres of basins, typical annual recharge 200,000 AF, Annual injection typically 15,000-30,000 AF	surface recharge injection	earliest operations 1917, barrier operations begun in 1970's

(continued)

Table 1.2 (continued)

agency	uses	designation	capacities	recharge method	comments
Mojave Water Agency	municipal agricultural	storage	76,000 AF annual entitlement for recharge	surface recharge	initiated in 1990's
Orange County Water District	municipal seawater intrusion	storage	1600 acres of basins typical annual recharge 300,000 AF injection 25,000 AF annual average	surface recharge injection	district created in 1933
Rosedale-Rio Bravo Storage District	agricultural	conjunctive use	390 acres of basins est. annual recharge capacity 145,000 AF	surface recharge	initiated in 1961
Santa Clara Valley Water District	municipal	conjunctive use	390 acres of basins typical annual recharge 100,000 AF	surface recharge	district created in 1929
Semitropic Water Storage District	municipal	banking	1,000,000 AF storage	surface recharge	district created in 1958
United Water Conservation District	municipal agricultural seawater intrusion	storage	400 acres of basins	surface recharge	facility construction in 1955
Zone 7 Water Agency	municipal	storage	target storage 225,000 AF	surface recharge	recharge begun in 1962

*source: Jones (2003)
abbreviations: ASR (aquifer storage and recovery), MWDSC (Metropolitan Water District of Southern California)

There are several instances where evolution of chemical composition of recharge waters has been observed during ASR tests or operation. In Charleston, SC, the composition of water injected into a limestone aquifer was observe to evolve from a Na-Ca-HCO_3 type to a Na-Cl type over test storage periods of 1 or 6 days (Mirecki et al., 1998). Recovery efficiencies ranged from 38-61% and were limited by excedence of the secondary drinking water standard for Cl. Input of chloride was attributed to upconing of porewater from a lower confining unit.

In Wessex, England, ASR tests using a single injection/extraction well were conducted in a Chalk aquifer where ambient groundwater was non-potable due to high fluoride concentrations. Elevated fluoride concentrations were also found in the recovered water and were attributed primarily to diffusive mixing but also to slow dissolution of the fluoride-containing mineral fluorite (Gaus et al., 2002).

In southwest Florida; water quality changes have been observed during ASR cycle tests (Arthur et al., 2002; Elison, 2003). Occurrence of arsenic and uranium in the recovered water at concentrations exceeding those in either the recharge water or ambient groundwater was attributed to release of arsenic and uranium from aquifer minerals. Contact between oxygenated recharge water and finely-disseminated pyrite was proposed as the basis for arsenic release (Arthur et al., 2002).

Chromium Accumulation in Groundwater

Silicate phases that contain Cr at trace levels may serve as a source of Cr in groundwater. Since these minerals contain Cr in the reduced form, their weathering will result in the slow release of Cr(III) into the dissolved phase (Bartlett, 1991). This process represents one mechanism for Cr(VI) accumulation in groundwater if the oxidation of Cr(III) to Cr(VI) is assumed to occur subsequent to (and more rapidly than) mineral dissolution. Since silicate mineral dissolution can be much slower than the water flow velocities commonly encountered, equilibrium assumptions do not apply and the kinetics of Cr release must, therefore, be considered.

Mineral Dissolution Kinetics

In general, for heterogeneous kinetics far from equilibrium, the dissolution or alteration rate of a mineral is written as (Lasaga, 1984):

$$\frac{dC_i}{dt} = \frac{A_m}{V} k_{im} \tag{1-1}$$

where dC_i/dt refers to the change in concentration of species i (C_i) due to mineral m, A_m is the surface area of mineral m, V is the volume of solution in contact with this surface, and k_{im} is the dissolution rate constant of species i from mineral m. In the absence of precipitation, dissolution of minerals is generally congruent, in which case the rate of dissolution of species i can be written as:

$$k_{im} = \nu_i k_m \tag{1-2}$$

where ν_i is the stoichiometric content of i in mineral m, and k_m is the overall dissolution rate constant for mineral m. This dissolution rate constant depends mostly on pH and temperature and to a lesser extent on the concentration of other species in solution (Lasaga, 1984).

Dissolution rate constants of silicate minerals reported in the literature span a range of several orders of magnitude depending on the specific minerals studied, and whether the studies were carried out in the laboratory or in the field. Differences between lab- and field-derived parameters have been attributed to different degrees of heterogeneity in the mineral surface and hydrologic factors influencing the effective surface area (Swoboda-Colberg and Drever, 1993; Velbel, 1993). Moreover, the use of different methods to estimate surface area introduces further uncertainty (White et al., 1996). Rate constant values gleaned from the literature on silicate weathering (converted to) are summarized in Table 1.3.

Possible Range of Chromium Release Rates

The results from previous research suggest that weathering rates of silicate minerals are extremely low, ranging from 10^{-17} to 10^{-11} mol $m^{-2}s^{-1}$. Given that the stoichiometric content of chromium in most silicate minerals is expected to relatively low (< 1%), congruent chromium release rates may be two or more orders of magnitude lower than these values.

The occurrence and distribution of Cr species in groundwater will depend both on the rate of Cr release through weathering reactions and the groundwater residence time. In ASR and particularly for small-scale systems, the retention time of the recharge water may be sufficiently short (relative to the rate of Cr release) that Cr concentrations in the recovered water may not exceed those in the recharge water.

Table 1.3
Silicate dissolution rate constants reported in the literature*

mineral	rate constant ($mol\ m^{-2}s^{-1}$)	pH	reference
plagioclase (oligoclase)	$\sim 1 \times 10^{-14}$	5.8-6.9	(Paces, 1983)
plagioclase (oligoclase)	1.2×10^{-16}		(White et al., 1996)
plagioclase (oligoclase)	4.0×10^{-14}		(White et al., 1996)
plagioclase	$(1.4\text{-}2.4) \times 10^{-11}$	4-4.5	(Swoboda-Colberg and Drever, 1993)
plagioclase	$(1.5\text{-}5.2) \times 10^{-14}$	4-4.5	(Swoboda-Colberg and Drever, 1993)
K-feldspar	$(1.7\text{-}2.1) \times 10^{-11}$	4-4.5	(Swoboda-Colberg and Drever, 1993)
K-feldspar	$(0.5\text{-}1.5) \times 10^{-13}$	4-4.5	(Swoboda-Colberg and Drever, 1993)
K-feldspar (anorthoclase)	3.2×10^{-17}		(White et al., 1996)
K-feldspar (anorthoclase)	1.6×10^{-14}		(White et al., 1996)
Muscovite	$(6.3\text{-}7.8) \times 10^{-12}$	4-4.5	(Swoboda-Colberg and Drever, 1993)
Muscovite	$(2.1\text{-}5.8) \times 10^{-14}$	4-4.5	(Swoboda-Colberg and Drever, 1993)
Biotite	3.8×10^{-12}	4.5	(Swoboda-Colberg and Drever, 1993)
Biotite	$(0.9\text{-}2.6) \times 10^{-14}$	4-4.5	(Swoboda-Colberg and Drever, 1993)
Chlorite	$(3.3\text{-}6.7) \times 10^{-12}$	4-4.5	(Swoboda-Colberg and Drever, 1993)
Chlorite	$(1\text{-}3) \times 10^{-14}$	4-4.5	(Swoboda-Colberg and Drever, 1993)
Hornblende	7.9×10^{-17}		(White et al., 1996)
Hornblende	3.2×10^{-14}		(White et al., 1996)
Hornblende	$(1.4\text{-}1.8) \times 10^{-12}$	4-4.5	(Swoboda-Colberg and Drever, 1993)
Hornblende	$(3\text{-}9) \times 10^{-15}$	4-4.5	(Swoboda-Colberg and Drever, 1993)
Enstatite**	$A\exp\left(\frac{E_a}{RT}\right)\left(\frac{a^2_{H^+}}{a_{Mg^{2+}}}\right)^{1/8}$		(Oelkers and Schott, 2001)

* converted to $mol\ m^{-2}\ s^{-1}$ where necessary

** $A = 2.4\ mol\ m^{-2}\ s^{-1}$, $E_a = 48.5\ kJ\ mol^{-1}$

CHAPTER 2
MATERIALS AND METHODS FOR EXPERIMENTAL STUDIES

REAGENTS

All chemicals were reagent grade or better and were used without further purification; $CaSO_4 \cdot 2H_2O$, CH_3COOH (HAc), CrO_3, $CrCl_3 \cdot 6H_2O$, EDTA disodium salt dehydrate ($C_{10}H_{14}N_2O_8Na_2 \cdot 2H_2O$), $FeCl_3 \cdot 6H_2O$, $Fe(NH_4)_2(SO_4)_2 \cdot 6H_2O$, HCl, $NaHCO_3$, $NaNO_3$, NaOH, Na_2SO_4, $NH_2OH \cdot HCl$ (HAH), and piperazine-N,N'-bis (2-ethanesulfonic acid) (PIPES) were obtained from EM Science. $Ca(NO_3)_2 \cdot 4H_2O$ (Fisher), DEAE Sephadex A-25 (40-125 μm, Sigma Chemical Co.), HNO_3 and NH_4OH (J.T. Baker), $Na_2B_4O_7 \cdot 10H_2O$ (Mallinckrodt), and NaOCl (6%, VWR) were also used. Solutions were prepared with water purified using a Millipore Milli-Q 18 MΩ system (Milli-Q water). Stock solutions of Cr(III), Fe(II), and Fe(III) were prepared by dissolving $CrCl_3 \cdot 6H_2O$, $Fe(NH_4)_2(SO_4)_2 \cdot 6H_2O$, and $FeCl_3 \cdot 6H_2O$, respectively, in 0.01 M HNO_3 (trace metal grade). Cr(VI) stock solutions were prepared by dissolving CrO_3 in Milli-Q water adjusted to pH 8.4 using 0.01 M NaOH.

SAMPLE COLLECTION, HANDLING AND PROCESSING

Types of Sample Containers and Cleaning

Laboratory experiments were conducted using reactors of different types and sizes depending on the volume of reaction medium. For experiments on redox-assisted coagulation, 1-L wide-mouth high-density polyethylene (HDPE) jars (VWR) or 1-L polypropylene beakers (Nalgene) were used. For other batch dissolution and sorption experiments, 150-, 250-, or 600-mL polypropylene beakers (Nalgene) were used as reactors. Two polymethylmethacrylate stirred flow-through reactors were used for flow-through dissolution of fuchsite. All the reactors used in the experiments were soaked in 4 M HNO_3 for at least 2 d and thoroughly washed with Milli-Q water, and dried in air before use. A 30-mL disposable plastic syringe (Sterile, VWR) was used as a resin cartridge for Cr speciation study. Fused silica wool (VWR) was soaked in 1 % HNO_3 for 2 d and rinsed with Milli-Q water before used as a bead support for the resin.

Solution samples were collected using pipettes or 1- or 5-mL disposable plastic syringes (Sterile, VWR) during the experiments and were stored in 15-mL or 50-mL disposable polypropylene centrifuge tubes (VWR). All disposable containers or syringes were used without further cleaning.

The sampling apparatus (bottle, cap, and tubing) used in the field were soaked in 4 M HNO_3 for 2 d, thoroughly washed with Milli-Q water, and dried in air before sampling trip. They were stored in clean plastic bags and transported to the field.

Sample Filtration

Water samples collected from field and laboratory were filtered through 0.2-μm, 0.02-μm, and/or ultra-fine filters. For 0.2-μm filtration, disposable syringe filters of polycarbonate film (Isopore™, Millipore) or cellulose acetate membrane (ValuPrep, VWR) were used. In addition, 0.2-μm cellulose nitrate membrane filters (47 mm in diameter, Type WCN, Whatman) were used to filter the effluent of flow-through reactors. For 0.02-μm filtration, solutions were filtered through an Anopore membrane disc (Anodisc™, Whatman), which are peripherally bonded to an annular polypropylene ring, using filter holders (Swin-Lok plastic filter holder, Corning) attached to 5- or 10-mL disposable plastic syringes (Sterile, VWR). For ultra-fine filtration were used Amicon Type YM-10 Centriplus membrane cones (Millipore) (referred to herein as an ultrafilter) with an effective molecular weight cutoff of 10,000 and an effective pore-size of approximately 0.003 μm. The ultrafilters were pretreated by rinsing with Milli-Q water and then with a small aliquot of each sample to saturate any possible adsorption sites on the filters and filtration containers (Rai et al., 1987).

Field Sample Collection and Processing

Water samples were collected in the field using a hand-operated vacuum pump (Nalgene) connected with clear plastic vacuum tubing (Nalgene) to transfer 1 L of water into a 1-L high density polyethylene (HDPE) bottle with a polypropylene filling/venting cap (Nalgene). The water samples were transported back to the laboratory (Caltech) and split into three subsamples (unfiltered, 0.2-μm filtered, and ultra-filtered). After processing, all sub-samples were acidified to a final concentration of 1% (v/v) with trace metal grade HNO_3. Acidified samples were analyzed for total Cr.

ANALYTICAL METHODS

Total dissolved Cr, Al, Si, and Fe concentrations were determined by inductively coupled plasma-mass spectrometry (ICP-MS, Hewlett Packard Model 4500). The minor isotopes ^{53}Cr and ^{57}Fe were measured to avoid isobaric interferences between ^{52}Cr and ArC and between ^{56}Fe and ArO. Detection limits (determined as three times the standard deviation of the blank) were 0.5 μg/L for Cr, 30 μg/L for Al and Si, and 5 μg/L for Fe. Samples were acidified to a final concentration of 1% (v/v) with trace metal grade HNO_3 for ICP-MS analysis. Calibration standard solutions were prepared by diluting commercially available single-element ICP standards (Ultra Scientific for Al and Si; EM Science for Cr and Fe).

In addition, Cr(VI) concentrations of selected samples were determined by ion chromatography (Dionex DX500) using a slightly modified version of EPA Method 218.6 (Arar et al., 1991). In this method, Cr(VI) stock solution (1000 mg/L) was prepared by dissolving K_2CrO_4 in Milli-Q water, rather than using $Na_2CrO_4 \cdot 4H_2O$ in ASTM Type 1 water. Samples were analyzed for Cr(VI) within 24 h at the Weymouth treatment plant of the Metropolitan Water District of Southern California (MWDSC).

Measurements of pH were made with a pH meter (VWR Model 8015) and a Ross glass electrode calibrated with pH 4.0, 7.0 and 10.0 buffers. In the field, the pH of water samples was measured using a portable pH meter (Model pH-40, Corning) calibrated with pH 7.0 and 10.0 buffers.

EXPERIMENTAL PROTOCOLS

Treatability Studies

Redox-assisted Coagulation

Since experimental methods have been described in detail (Lee and Hering, 2003), only a brief summary is provided here. Experiments were conducted to examine the removal of Cr from solution under varying conditions. Parameters examined included Cr and Fe oxidation states, pH and solution composition, and reaction time. Three reaction solutions used were (1) Synthetic CRW (0.01 M $NaNO_3$, 2 mM $NaHCO_3$, and 2 mM $Ca(NO_3)_2$), (2) Natural CRW (raw water obtained from the MWDSC Weymouth treatment plant), and (3) Simple electrolyte (0.01 M $NaNO_3$ and 1 mM PIPES). Experiments were conducted at room temperature (23.0 ± 0.4 °C) using 1-L polyethylene reactors that were open to the atmosphere.

The general procedure for all experiments was as follows. Reaction solutions were adjusted (if necessary) to the desired pH and spiked with Cr at varying concentrations (≤ 1 μM) as either Cr(VI) or Cr(III) and mixed on a magnetic stirrer at 100 rpm. Subsequently, the solutions were spiked with Fe(II) to a final concentration of up to 30 μM. Solutions were then immediately amended with appropriate volumes of 0.1 M NaOH to maintain the initial pH against the proton release associated with iron hydrolysis. The suspensions were mixed at 100 rpm for 30 min and, after a settling time of 15 min, samples were collected using a polyethylene syringe and (for experiments with synthetic CRW) filtered using either 0.2-μm filters or ultrafilters. Since the total dissolved Cr and Fe concentrations in 0.2-μm and ultrafilter filtrates were identical for synthetic CRW samples, the aliquots of the natural CRW and the simple electrolyte were filtered only using 0.2-μm filters.

Filterability of colloidal Cr(III)

The filterability of colloidal Cr(III) was examined using a 1-L solution of 0.01 M $NaNO_3$ spiked with 1.00 μM Cr(III) without any coagulant. The solution was supersaturated with respect to amorphous $Cr(OH)_{3(s)}$ by raising the pH to 8.2. The aliquots of the solution were filtered using 0.2-μm and 0.02-μm filters, and ultrafilters.

Kinetics

Redox kinetics between Cr(VI) and Fe(II) in the presence of O_2 were examined at pH 6.5 and 7.0 using simple electrolyte solution. The pH was monitored during the experiments and was constant to within ± 0.01 units. Ten 5-mL samples of each aliquot were collected at time intervals of 1 - 10 min after the addition of Fe(II) stock solution and filtered with a 0.2 μm filter.

Removal of Cr(VI) with Fe(III)

The efficiency of Cr(VI) removal from synthetic CRW by coagulation with $FeCl_3$ was examined at high ratios of total Fe to total Cr (Fe_T:Cr_T = r = 50 - 200) in the presence and absence of 3 mM Na_2SO_4. During these experiments solutions were mixed at 45 rpm for 30 min after the addition of Fe(III).

The extent of Cr(VI) sorption onto freshly precipitated $Fe(OH)_{3(am)}$ was also examined at a low Fe_T:Cr_T ratio (r = 5). Aliquots of synthetic CRW and simple electrolyte were first amended with an appropriate volume of Fe(III) stock solution and the pH was adjusted to 8.2. Then, 1.00 μM Cr(VI) was spiked into the solutions. The pH was then decreased stepwise by the addition of 0.1 M HNO_3 to a final pH of ~5.0. At each step, the pH was stabilized for 5 – 10 min before samples were collected.

Weathering Studies

Solid Phases

Brazilian fuchsite, obtained from the Smithsonian Institution, Washington, D.C., synthetic $Cr(OH)_3(s)$, and aquifer sediments from the Cadiz area were used in this study. Fuchsite is a Cr-rich variety of muscovite. Fuchsite specimens were green flaky aggregates and were typically 5 × 7 × 2 cm in dimension. Six specimens were crushed and ground in a ceramic mortar and were sieved into a size fraction of 125 - 88 μm. Each of the size-fractionated samples was washed several times with Milli-Q water and was subsequently cleaned of ultrafine particles by repeated (~10 times) ultrasonic cleaning in acetone until a clear supernatant was obtained. The cleaned powders were dried overnight in an oven at 60 °C and were stored in glass vials.

Aquifer sediment samples were obtained as drill-cuttings (OW-2) and provided by Cadiz Inc. The sample provided was completely dried and had been stored in plastic bags for a long period in a trailer near Cadiz Ranch. About 200 g of the sample was brought to the laboratory (Caltech) and was crushed and sieved to collect 125 - 88 μm fraction. The surface area of the fraction was measured by BET-N_2 adsorption using a Gemini 2360 surface area analyzer without washing and was 5.22 m^2/g.

Elemental compositions of the fuchsite and Cadiz samples were determined by X-ray fluorescence spectrometry (XRF) on a fused disc prepared from 2 g of cleaned fuchsite powders at XRAL Laboratories, Ontario, Canada. The analytical detection limit was 0.01% for the constituents. Powdered fuchsite samples were analyzed by X-ray diffraction (XRD) to determine if any other minerals were incorporated as impurities. The XRD analysis was conducted on a Scintag Pad V X-ray Powder Diffractometer using Cu Kα radiation on a theta-2 theta goniometer equipped with a germanium solid-state detector. Acquisition conditions were 35 KV and 30 mA. Scans were obtained typically from 5 to 70° 2θ, with step size of 0.02° 2θ, with a count time of 1 sec. The X-ray diffractograms of the fuchsite samples were all identical to that of pure muscovite and no impurities were detected. Surface area of the fuchsite samples was determined by BET N_2-adsorption. The surface area was 0.81 ± 0.33 m^2/g of fuchsite (the fraction of 125 - 88 μm) based on twenty nine measurements.

Amorphous $Cr(OH)_3(s)$ was synthesized by adding 200 mL of chromium(III) stock solution (0.3 M Cr(III) in 0.1 M HNO_3) to 800 mL of Milli-Q water. The pH of the solution was adjusted to 9.5 with 1.0 M NaOH to precipitate $Cr(OH)_{3(s)}$. The suspension was aged for 4 hr, and the supernatant was discarded after centrifugation. The $Cr(OH)_3$ precipitates were then rinsed 5 times with Milli-Q water and dried overnight in an oven at 60 °C. The surface area of $Cr(OH)_3(s)$ was 215.0 ± 1.1 m^2/g.

Batch Dissolution Experiments

Batch dissolution experiments were conducted using fuchsite, synthetic $Cr(OH)_{3(s)}$, or aquifer sediment samples at pH 9 and 2 under oxidizing and non-oxidizing conditions. Sodium hypochlorite (1 mM) was used as an oxidant for oxidizing dissolution media. The pH of each medium was first adjusted to a desired value (2.0 or 9.0) using 0.1 M HNO_3 or NaOH. Subsequently, each run was initiated by adding an aliquot of the solids to a dissolution medium in a reactor (250-mL polypropylene beaker). During the course of the dissolution experiment, the solution was moderately stirred with a Teflon stir bar and a magnetic stirring plate, and the reactor was covered with parafilm to minimize evaporation.

For fuchsite dissolution (Table 2.1), the pH was maintained within ± 0.05 unit throughout the experiments by adding 0.1 M HNO_3 or NaOH using an autotitrator (METTLER TOLED DL 50 Graphix Autotitrator). Temperature of the suspension was initially 25.0 °C and slowly increased up to a maximum of 27.2 °C with an average of 26.8 ± 0.7 °C due to the heat generated by the magnetic stirring plate. Three-mL of sample was taken from the reactor over time using 1-mL polypropylene syringe and was filtered using 0.2-μm filters attached to the syringe. The first 1 mL of filtrate was discarded and the rest was diluted appropriately with 1 % HNO_3 for measurement of dissolved concentrations of Cr, Al, Si, and Fe by ICP-MS.

For $Cr(OH)_{3(s)}$ dissolution (Table 2.2), the pH was constant within ± 0.12 unit without using an autotitrator. The average temperature of the suspension was 25.6 ± 0.8 °C. Samples were taken by filtering 5 mL of suspension through 0.2-μm filter, and the last 3 mL of filtrate was diluted with 1 % HNO_3 for measurement of dissolved Cr concentrations. Filterability of $Cr(OH)_{3(s)}$ was examined with a suspension of the solid (1.0 g/L) prepared in 0.01 M $NaNO_3$. Aliquots of the suspension were immediately filtered using either a 0.2-μm filter or an ultrafilter. The total dissolved Cr concentrations in the filtrates of 0.2-μm and ultrafilters were both < 0.5 μg/L, indicating $Cr(OH)_{3(s)}$ was effectively filtered out using 0.2-μm filter.

The dissolution of aquifer sediment samples was conducted only at pH 9 under oxidizing and non-oxidizing conditions using borate buffer instead of the autotitrator. Generally, the procedure and the sample collection were identical to those of the fuchsite dissolution at pH 9. However, for non-oxidizing conditions, a higher concentration (500 μM) of EDTA was used to stabilize dissolved metals. A preliminary dissolution experiment was run, in which the dry solid was suspended in buffer and immediately exposed to the oxidant. In a subsequent experiment, the sample was pre-equilibrated with borate buffer solution (0.01 M $NaNO_3$, 0.01 M $Na_2B_4O_7$) for two days. After the pre-equilibration, the solid was washed 5 times with the clean buffer solution before oxidative dissolution was initiated by adding 1 mM NaOCl. The solid concentration was 10.0 g/L and the pH was 9.02 ± 0.06 and 9.08 ± 0.08 under non-oxidizing and oxidizing conditions, respectively.

Table 2.1
Experimental conditions for batch dissolution of fuchsite

experiment ID*	reaction medium	pH	sample ID	surface area (m^2/g)	solid conc. (g/L)	reaction time (h)
BDF2NOx	0.01 M $NaNO_3$	1.99 ± 0.02	F4	1.199	4.00	359
BDF2Ox	0.01 M $NaNO_3$, 1 mM NaOCl	2.00 ± 0.02	F4	1.199	4.00	432
BDF9NOx	0.01 M $NaNO_3$	9.02 ± 0.08	F2	0.957	10.54	240
BDF9NOxE	0.01 M $NaNO_3$, 50 μM EDTA	9.03 ± 0.08	F2	1.105	10.74	362
BDF9Ox	0.01 M $NaNO_3$, 1 mM NaOCl	9.01 ± 0.04	F1	0.890	4.01	312

*Experiment ID is designated as follows: (BD) batch dissolution, (F) fuchsite, (NOx) non-oxidizing condition, (NOxE) non-oxidizing condition in the presence of EDTA, (Ox) oxidizing condition, and numbers in the ID indicate the pH of reaction medium.

Table 2.2
Experimental conditions for batch dissolution of $Cr(OH)_3(s)$

experiment ID*	reaction medium	pH	surface area (m^2/g)	solid conc. (g/L)	reaction time (h)
BDCr2NOx	0.01 M $NaNO_3$	1.99 ± 0.01	213.8	1.00	7.1
BDCr2Ox	0.01 M $NaNO_3$ + 20 mM NaOCl	2.16 ± 0.07	216.3	1.00	3.0
BDCr9NOxE	0.01 M $NaNO_3$ + 100 μM EDTA	8.99 ± 0.01	215.2	1.00	7.1
BDCr9Ox	0.01 M $NaNO_3$ + 20 mM NaOCl	8.86 ± 0.12	214.6	1.00	3.0

*Experiment ID as in Table 2.1 except: (Cr) $Cr(OH)_3(s)$

Table 2.3
Experimental conditions for flow-through dissolution of fuchsite

experiment ID†	influent solution*	pH	sample ID	surface area (m^2/g)	solid conc. (g/L)	flow rate (mL/h)	reaction time (h)
FTDF2NOx	(no addition)	2.00 ± 0.02	F1	0.890	18.28	1.47 ± 0.08	473
FTDF2Ox	1 mM NaOCl	2.01 ± 0.02	F1	0.890	18.45	1.68 ± 0.10	473
FTDF9NoxE	0.01 M $Na_2B_4O_7$, 200 μM EDTA	9.09 ± 0.03	F6	0.906	19.17	1.48 ± 0.31	356
FTDF9Ox	0.01 M $Na_2B_4O_7$, 1 mM NaOCl	9.10 ± 0.03	F6	0.906	19.35	1.77 ± 0.11	356

† Experiment ID as in Table 2.1 except: (FTD) flow-through dissolution *All influent solutions contained 0.01 M $NaNO_3$

Flow-through Dissolution Experiments

Experiments were conducted using two polymethylmethacrylate stirred flow-through reactors (~50 mL) as described previously (Giammar and Hering, 2002; Kraemer, 1997). Experiments were conducted at the ambient temperature of the laboratory (23.0 ± 0.4 °C). Two reactors were run parallel at the same pH (9 or 2) for oxidizing and non-oxidizing conditions (scc Table 2.3 for details). Flow was controlled by a peristaltic pump and stirring was provided by a Teflon stir bar and a magnetic stirring plate. The effluent was passed through double 0.2-μm filters attached to a 47-mm polypropylene filter holder capping the reactor. Total dissolved Cr, Al, Si and Fe concentrations in the effluent were determined by ICP-MS.

The steady-state release rate, R ($mol \cdot m^{-2} s^{-1}$), of each element from the solid was calculated using the following equation:

$$R = \frac{1}{S \cdot A} \cdot \frac{d[M]_{diss}}{dt} = \frac{1}{S \cdot A} \cdot [M]_{diss} \cdot \frac{q}{3600 \cdot V} \quad (2.1)$$

where $[M]_{diss}$ is total dissolved metal concentrations [$mol \cdot L^{-1}$] in the effluent, S is the solid concentration [$g \cdot L^{-1}$], A is the specific surface area of the solid [$m^2 \cdot g^{-1}$], q is the flow rate [$mL \cdot h^{-1}$] and V is the reactor volume [mL]. Steady-state release rates were determined after metal concentrations in the effluent reached a constant value.

Chromate Sorption Experiments

Sorption of Cr(VI) on fuchsite (125 - 88 μm fraction) or on $Cr(OH)_3$ was determined in batch systems as a function of both the pH and variable sorbate concentration. All experiments were conducted with a background electrolyte of 0.01 M $NaNO_3$ at 25.5 ± 0.7 °C. For pH experiments, the solid was added to a solution of Cr(VI) in 0.01 M $NaNO_3$ at pH ~ 7. Subsequently, the suspension was magnetically stirred and the pH was lowered stepwise by addition of 0.1 or 1.0 M HNO_3 to a final pH of 2. At each step, a sample was taken from the suspension 5 - 15 min after the pH stabilized.

The sorption isotherm for Cr(VI) was obtained at pH ~2 for fuchsite (Table 2.4) and at pH ~3 for $Cr(OH)_3(s)$ (Table 2.5). Aliquots of Cr(VI) stock solutions were added to the solid suspensions to yield the desired final concentrations of $[Cr(VI)]_T$. The suspensions were then equilibrated for 30 min (fuchsite) or 6 h ($Cr(OH)_3(s)$). The short reaction time was chosen for fuchsite in order to minimize dissolution of the sorbent because the amounts of Cr released from fuchsite could be larger than those of sorbed Cr with an extended reaction time. All samples were filtered through 0.2-μm filters and were acidified to 1 % HNO_3 for measurement of total dissolved metal concentrations using ICP-MS.

Table 2.4
Experimental conditions* for chromate sorption on fuchsite

experiment type	sample ID	$[Cr(VI)]_T$ (μM)	solid conc. (g/L)	pH range or pH	surface area (m^2/g)	equilibration time (min)
pH edge	F4	1.93	4.01	6.88 - 1.82	1.199	5 - 10
	F1	0.41	10.29	4.40 - 1.85	0.890	360
isotherm	F1	0 - 1.48	20.04 ± 0.03	1.98 ± 0.06	0.890	30
	F4	0 - 8.94	10.02 ± 0.02	2.00 ± 0.01	1.199	30

* All experiments conducted in a background electrolyte of 0.01 M $NaNO_3$

Table 2.5
Experimental conditions* for chromate sorption on $Cr(OH)_3(s)$

experiment type	$[Cr(VI)]_T$ (μM)	solids conc. (g/L)	pH range or pH	surface area (m^2/g)	equilibration time (min)
pH edge	177.5	1.00	7.59 - 2.00	215.0 ± 1.1	5 - 15
Isotherm	0 - 89.3	0.05	2.95 ± 0.02		360

* All experiments conducted in a background electrolyte of 0.01 M $NaNO_3$

Chromium Speciation Studies

Speciation procedure followed the procedure proposed by Beaubien et al. (1994) with a smaller volume of samples (10 mL) instead of 500 mL samples as used in the original work. DEAE resin was soaked in 2 M HCl for 24 h, then for 3 d in 0.1 M HCl and rinsed thoroughly with Milli-Q water. After cleaning, the slurry of the resin in Milli-Q water was adjusted to pH 5.8 - 6.0 by adding 3.0 M NH_4OH. A 30-mL plastic syringe was modified to act as resin cartridge and the resin (1.35 g dry weight) was supported by silica wool in the cartridge. Flow rate of samples through the resin was measured each time and was constant at 8.6 mL/min. Cr(III) and Cr(VI) sample solutions were prepared in triplicate by diluting Cr(III) and C(VI) stock solutions, respectively. Cr(III) sample was prepared as 0.75 μM (39.3 μg/L) of Cr(III) in 0.01 M $NaNO_3$ at pH 2.92 and Cr(VI) sample as 0.76 μM (39.7 μg/L) in 0.01 M $NaNO_3$ at pH 8.03.

For the separation of Cr species, 0.02 mL of 7.5 M acetic acid (HAc) was added into each replicate sample (10 mL) to buffer the solution at pH 4.7. Subsequently, a sample was allowed to pass through the resin at a constant rate, collected in a 50-mL polypropylene centrifuge tube, and acidified by adding 0.2 mL of 50 % HNO_3 for analysis of the fraction of dissolved Cr(III).

Once the entire sample passed through the cartridge, the cartridge was vacuum drained and 1.5 mL of 5 % hydroxylamine hydrochloride (HAH) was added to the resin to reductively elute the resin-bound Cr(VI). After soaking for 10 min, the cartridge was vacuum-drained and rinsed with 1.5 mL of Milli-Q water, which was collected in a sample collector (50-mL centrifuge tube). This procedure was repeated two more times and then 1 mL of Milli-Q water was added into the sample collector to make a final volume of 10 mL, which was acidified for analysis of the fraction of dissolved Cr(VI).

After the Cr(VI) elution, the resin was soaked with 1.5 mL of 4 M HNO_3 for 10 min, vacuum-drained, and rinsed with equal volume of Milli-Q water, which was repeated one more time. Finally, the resin was rinsed with 4 mL of Milli-Q water, which was added into the sample collector to make a final volume of 10 mL, which represented the fraction of colloidal Cr from the original sample. Calibration standard solutions for ICP-MS analyses were prepared with the same matrix of the fraction of each Cr species.

MODELING

Treatability Studies

Rate expressions and constants previously reported for the reduction of Cr(VI) by Fe(II) and for oxidation of Fe(II) by O_2 were used to predict the rate and extent of Cr(VI) reduction and Fe(II) oxidation under our experimental conditions. Simulations were performed using SCIENTIST (MicroMath®) considering the reactions of Cr(VI) and Fe(II). The reader is referred to the published work for the detailed rate expressions (Lee and Hering, 2003).

The reaction of Fe(II) with O_2 was explicitly considered in modeling our experiments. Furthermore, it was assumed that all Cr(III) produced by reduction of Cr(VI) and all Fe(III) produced by oxidation of Fe(II) were quantitatively removed by filtration. Thus, the fraction of Cr(VI) removed was taken to correspond to the fraction of Cr(VI) reduced and the fraction of Fe(II) removed to the fraction of Fe(II) oxidized.

Modeling of Cr(VI) sorption onto freshly precipitated $Fe(OH)_{3(am)}$ from the simple electrolyte was performed using MINEQL+ v. 4.5 (Schecher and McAvoy, 1998). Relevant thermodynamic data for aqueous species and the equilibrium constants for surface complexation reactions were used as given in MINEQL+ database; MINEQL+ constants for reactions at the hydrous ferric oxide surface are taken from Dzombak and Morel (1990).

Modeling Chemical Evolution of Groundwater

Simulations of dissolved concentrations over time were performed using SCIENTIST for batch reactor conditions for two cases: (1) dissolution of a primary solid containing constituent X and (2) oxidative dissolution of $Cr(OH)_3(s)$. In both cases, the initial solid was considered not to be depleted during the reaction. In the first case, dissolution was considered alone and in combination with precipitation of a secondary solid also containing X and/or adsorption of X onto a surface with limited site density. In the second case, dissolution was considered with adsorption of Cr(VI) (i.e., the product of the dissolution reaction). Kinetics of adsorption and desorption were explicitly included and inhibition of oxidative dissolution due to adsorption was considered.

CHAPTER 3
METHODS FOR NUMERICAL STUDIES

INTRODUCTION

This chapter describes the numerical models and approaches to investigating the potential impact of natural chromium sources on aquifer storage and recovery operations (ASR). The chapter begins with a brief overview of the computer codes used in this research and the governing partial differential equations solved by these codes. Following the description of the models, two approaches to modeling an ASR system are presented.

The first system is modeled using a single domain that encompasses both the unsaturated and saturated zones of the subsurface. Using the single domain approach, different scenarios accounting for dissolve chromium occurrence in groundwater are used to estimate the potential impact on the quality of clean water stored in the aquifer. The scenarios are based loosely on observations at the Cadiz site in the Fenner Gap, but are largely hypothetical and could apply to similar sites in the arid and semi-arid western United States. Within the framework of these scenarios, the sensitivity of the model to spatial and temporal discretization is first demonstrated. Then, for a sufficiently robust discretization scheme, a series of simulations testing the models' sensitivity to key flow and transport parameters are described.

A second modeling approach was investigated as a means of reducing the computational burden imposed by the single domain approach, while preserving the numerical accuracy of the simulation. This is a dual domain approach where an output water flux from the unsaturated domain is adapted to become the input for the saturated domain model.

GROUNDWATER FLOW AND TRANSPORT MODELS

Transport Code: FEMWATER

FEMWATER version 2.1 was used to simulate flow and transport in the saturated and unsaturated zone (Lin et al., 1997). The Department of Defense Groundwater Modeling System (GMS), developed by Engineering Computer Graphics Laboratory of Brigham Young University, was used as a input/output graphical interface. The results of a FEMWATER simulation include pressure head, moisture content, groundwater velocity and tracer concentration as a function of time and space on a user specified grid. All FEMWATER simulations were conducted with an Intel Pentium 4 processor, 2.5 GHz, 1.0 GB of RD RAM.

Reactive Transport Code: HBGC123D

For transport simulations with and without reaction in the saturated and unsaturated zone the computer code HydroBioGeoChem 123D (HBGC123D), version 1.1 (Gwo et al., 1999), was used. This is a public-domain code developed at the Oak Ridge National Laboratory, and is an extended version of HydroBioGeoChem (Yeh et al., 1998). HBGC123D solves hydrologic transport and biogeochemical kinetic and/or equilibrium reactions in one, two, or three-dimensions. For transient flow simulations this code requires as input the results of a flow model like FEMWATER. HBGC123D calculates concentration of all species as a function of time and space throughout the model domain. All HBGC123D simulations were conducted with an Intel Pentium 4 processor, 2.5 GHz, 1.0 GB of RD RAM.

Governing Equations for Flow (FEMWATER)

Flow is described with a modified version of the Richards equation:

$$\left(\frac{d\theta_w}{dh}\right)\frac{\partial h}{\partial t} = \nabla\left[\left(\frac{\rho g}{\mu}\mathbf{k}_s k_r\right)(\nabla h + \nabla z)\right] + q \tag{3-1}$$

where θ_w is the moisture content; h is the pressure head; t is time; ρ is the water density, g is the acceleration of gravity; μ is the dynamic viscosity of water; $\mathbf{k_s}$ is the saturated permeability tensor; z is the potential head; q is the source and/or sink; and k_r is the relative permeability. Within the unsaturated zone, relative permeability and water saturation are described using the van Genuchten equations (van Genuchten, 1980):

$$k_r = \theta_e^{0.5}\left[1-\left(1-\theta_e^{1/\gamma}\right)^{\gamma}\right]^2 \tag{3-2}$$

where,

$$\theta_e = \begin{cases} \left[1+\left(|\alpha h|\right)^{\beta}\right]^{-\gamma} & \text{for } h<0 \\ 1 & \text{for } h \ge 0 \end{cases} \tag{3-3}$$

$$\gamma = 1-\frac{1}{\beta} \tag{3-4}$$

and

$$\theta_w = \theta_r + \theta_e(\theta_s - \theta_r) \tag{3-5}$$

Where, θ_e is the effective moisture content, β and γ are soil-specific exponents, α is a soil-specific coefficient, θ_r is the residual moisture content, and θ_s is the saturation moisture content.

The governing equation for transport is derived from the continuity of mass and flux laws. The major processes are advection, dispersion/diffusion, and source/sink.

$$\theta_w\frac{\partial C}{\partial t} + \mathbf{V}\cdot\nabla C - \nabla\cdot\left[\left(a_T|\mathbf{V}|\delta + (a_L - a_T)\frac{\mathbf{VV}}{|\mathbf{V}|} + a_M\theta_w\tau\delta\right)\cdot\nabla C\right] = +qC_{in} - qC + \left(\frac{\partial\theta_w}{\partial h}\left(\frac{\partial h}{\partial t}\right) + \nabla\cdot\mathbf{V} - \frac{\partial\theta_w}{\partial t}\right)C \tag{3-6}$$

where C is the material concentration in the aqueous phase; δ is the Kronecker delta tensor; a_T is the lateral dispersivity; a_L is the longitudinal dispersivity; a_M is the molecular diffusion coefficient; τ is the tortuosity, C_{in} is the material concentration in the source, and **V** is the Darcy velocity. The Darcy velocity is calculated as follows

$$\mathbf{V} = -\left(\frac{\rho g}{\mu}\mathbf{k}_s k_r\right)(\nabla h + \nabla z) \tag{3-7}$$

The transport equation is solved employing a hybrid Lagrangian-Eulerian methodology. The inability of FEMWATER to solve geochemical reaction equations (e.g. a chromium-generating reaction) results in Cr being treated as a conservative constituent.

Governing Equations for Transport (FEMWATER and HBGC123D)

The governing equations for transport are derived based on the continuity of mass and flux laws. The major processes are advection, dispersion/diffusion, adsorption, decay, biodegradation, and source/sink.

$$\theta\frac{\partial C}{\partial t} + \rho_b\frac{\partial S}{\partial t} + V\cdot\nabla C - \nabla\cdot\left[\left(a_T|V|\delta + (a_L - a_T)\frac{VV}{|V|} + a_m\theta\tau\delta\right)\cdot\nabla C\right] =$$
$$-\left(\alpha'\frac{\partial h}{\partial t} + \lambda\right)(\theta C + \rho_b S) - (\theta K_W C + \rho_b K_s S) + m - \frac{\rho^*}{\rho}qC \tag{3-8}$$
$$+\left(F\frac{\partial h}{\partial t} + \frac{\rho_0}{\rho}V\cdot\nabla\left(\frac{\rho}{\rho_0}\right) - \frac{\partial\theta}{\partial t}\right)C$$

where θ is the moisture concentration; ρ_b is the bulk density of the medium; C is the material concentration in the aqueous phase; S is the material concentration in the adsorbed phase; t is time; V is discharge; ∇ is the del operator; α' is the compressibility of the medium; h is the pressure head; λ is the decay constant; m is the artificial mass rate; K_W is the first order biodegradation rate constant through the dissolved phase; K_s is the first order biodegradation rate through the adsorbed phase; F is the storage coefficient; K_d is the distribution coefficient; S_{max} is the maximum concentration of the medium in the Langmuir nonlinear isotherm; n is the power index in the Freundlich nonlinear isotherm; K is the coefficient in the Langmuir or Freundlich nonlinear isotherm; δ is the Kronecker delta tensor; a_T is the lateral dispersivity; a_L is the longitudinal dispersivity; a_m is the molecular diffusion coefficient; and τ is the tortuosity. The transport equation is solved employing a hybrid Lagrangian-Eulerian methodology. The inability of FEMWATER to solve geochemical reaction equations (e.g. a chromium-generating reaction) forces the modeler to treat Cr as a conservative constituent.

SINGLE DOMAIN FLOW AND TRANSPORT SIMULATIONS

Initial simulations were undertaken in a single domain that encompassed both unsaturated and saturated media. The simultaneous simulation of saturated and unsaturated flow in the same domain required a relatively fine finite element grid and imposed significant computational burden. In addition, the mounding created by the infiltration of artificially recharged water is highly sensitive to the unsaturated zone hydraulic parameters, and greatly impacts the vertical transfer of clean water into the saturated media below. While the array of single domain simulations was being executed and analyzed, a more elegant, less computationally burdensome method of splitting the unsaturated and saturated aspects of the simulation was investigated. The methods behind that investigation are discussed in section 3.4.

Three source scenarios were investigated that would be compatible with the water quality monitoring results during the pilot pump and recharge test.

Scenario I: This modeling scenario assumes that a steady concentration of a nonspecific Cr(VI) species from an unknown, up-gradient source arrives at Fenner gap with the regional inflow of groundwater. No Cr(VI) source is applied in the simulation domain, and the ASR water quality issue is governed by hydrodynamic mixing and dilution of the clean (input) water with the native, chromium-impacted groundwater.

Scenario II: This scenario simulates constant and uniform chromium dissolution as Cr(VI) from the alluvium throughout the groundwater basin, including the simulation domain. In this case, the Cr(VI) concentrations present in the pilot test area are observed as uniform because of the long travel time through the domain relative to the weather rates controlling the Cr(VI) release.

Scenario III: This scenario simulates constant and uniform chromium dissolution in the unsaturated portion of the single domain. In this case, Cr(VI) is allowed to enter the aquifer only by dissolution into the initially clean recharge water. The purpose of simulating this scenario was to determine the potential contribution of the unsaturated zone weathering in the ASR problem.

Single Domain Model

For all simulations the model domain was a rectangular box 2,160 m long, 1,500 m wide and 250 m deep (Figure 3.1), with the initial water table located at an elevation of approximately 180 m. A set of recharge wells was employed to model the spreading basin. The recharge wells were centered approximately 750 m away from the model entrance boundary. All simulations employed a homogeneous, isotropic medium. The value of hydraulic conductivity obtained from pumping tests in the pilot field test area was used (Table 3.1), and soil properties for the unsaturated zone were those listed in FEMWATER as average values for a sandy medium.

A 3-D mesh with irregularly sized elements was used to represent the model domain. The spatial resolution was increased vertically in the unsaturated zone and horizontally around the well pattern or spreading basin center. After a preliminary investigation (discussed in the next section) of several model boxes of these characteristics, one containing 20,615 nodes and 38,400 elements was selected for the rest of this work.

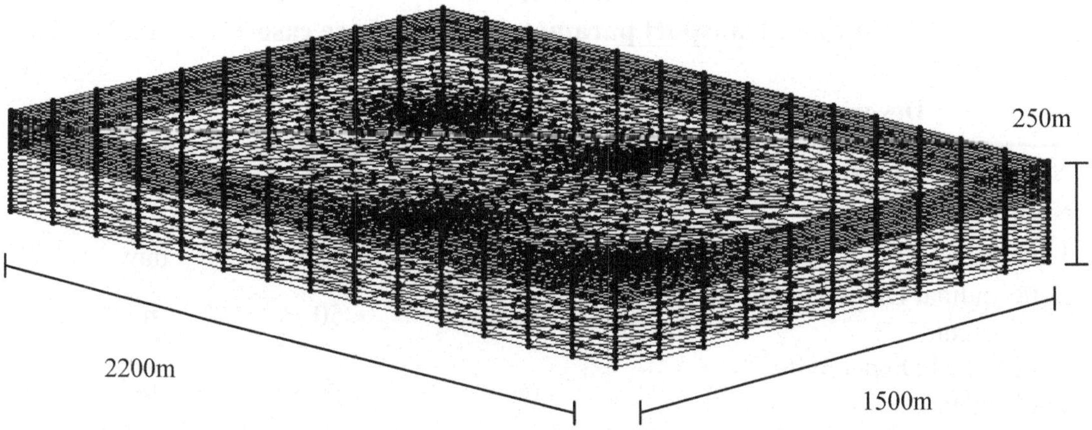

Figure 3.1 Three-dimension finite element grid for the groundwater modeling domain, where element density increases near injection and extraction zones (note: for separate plan and cross-sectional views see Figure 3.2).

Single Domain Base Case Simulations

One specific set of conditions was taken as the basis for comparison with all other simulations. In the base case, initial and boundary conditions were those used in Scenario I. Clean water was injected at five recharge well nodes located at an elevation of approximately 160 m (20 m under the water table) for a period of approximately 5 years (1,800 days) at a rate of 6,000 m^3/d (1,200 m^3/d at each well node). Space and time resolution were selected after a preliminary investigation. Longitudinal and transversal dispersivity values were set at 50 and 5 m, respectively, which are roughly applicable for a domain of this size (Gelhar et al., 1992). These and other flow and transport parameters are listed in Table 3.1.

Flow and transport were always solved separately: flow fields obtained with FEMWATER were fed to either FEMWATER (non-reactive case) or HBGC123D (non-reactive and reactive case) to conduct transport simulations.

Single Domain Model Initial and Boundary Conditions

Under Scenario I assumptions, clean water was introduced into the system and allowed to mix with native groundwater containing Cr at a uniform concentration. Model response to different aquifer dispersivity values, recharge rates and recharge configurations was observed and compared to the base case. Scenario II investigations included the impact of different rates of Cr dissolution from aquifer materials on the resulting concentration patterns when clean water was introduced under the flow conditions of the base case. Initial concentrations were those of steady state chromium dissolution with groundwater background flow. The results were compared at two different groundwater concentrations entering the model domain. Under Scenario III clean water infiltrated through the ground and allowed to collect Cr as it moved through the vadose zone. Dissolution rates for Scenario III were the same as those for Scenario II.

Table 3.1
Flow and transport parameters for the base case

Description	Symbol	Value	Units
Number of Nodes		20,615	-
Number of Elements		38,400	-
Time Step	Δt	100	day
Longitudinal Dispersivity Coefficient	α_L	50	m
Transversal Dispersivity Coefficient	α_T	5	m
Number of Recharge Nodes		5	-
Recharge Rate	Q_r	6,000	m^3/day
Length of Recharge Period		1,800	day
Bulk Density of the Medium	ρ_b	1,510	kg/m^3
Permeability	K	3.00E-11	m^2
Tortuosity	τ	0.7	-
Molecular Diffusion Coefficient	D	8.64E-05	m^2/day

Flow Model Initial and Boundary Conditions

On the right and left sides of the box, constant head (Dirichlet) boundary conditions were assumed (Figure 3.2), based on groundwater elevation contours measured in the study area and negligible drawdown observed at those distances from the well during the pump tests. Head values were set at 184.4 m and 180.7 m on the right and left boundaries respectively. This corresponds to a hydraulic gradient of 0.0017. The initial aquifer conditions represent the baseline flow with an inclined water table between the two boundaries. The front, back, and bottom of the box were set to impervious boundaries, simulating bedrock confinement. The top of the box was also set as a no flow boundary in all cases, except where recharge from the ground was simulated: here the spreading basin area was defined as a constant flux boundary condition. For all of the simulations, the quantity of clean water injected was kept constant at 1.08e7m^3. Recharge conditions are described in more detail under each scenario.

Transport Model Initial and Boundary Conditions

The variable boundary condition was used at the entrance and exit of the model domain for HBGC123D simulations. For FEMWATER runs, the variable boundary condition was also used at the entrance whereas the gradient flux was applied to the exit boundary.

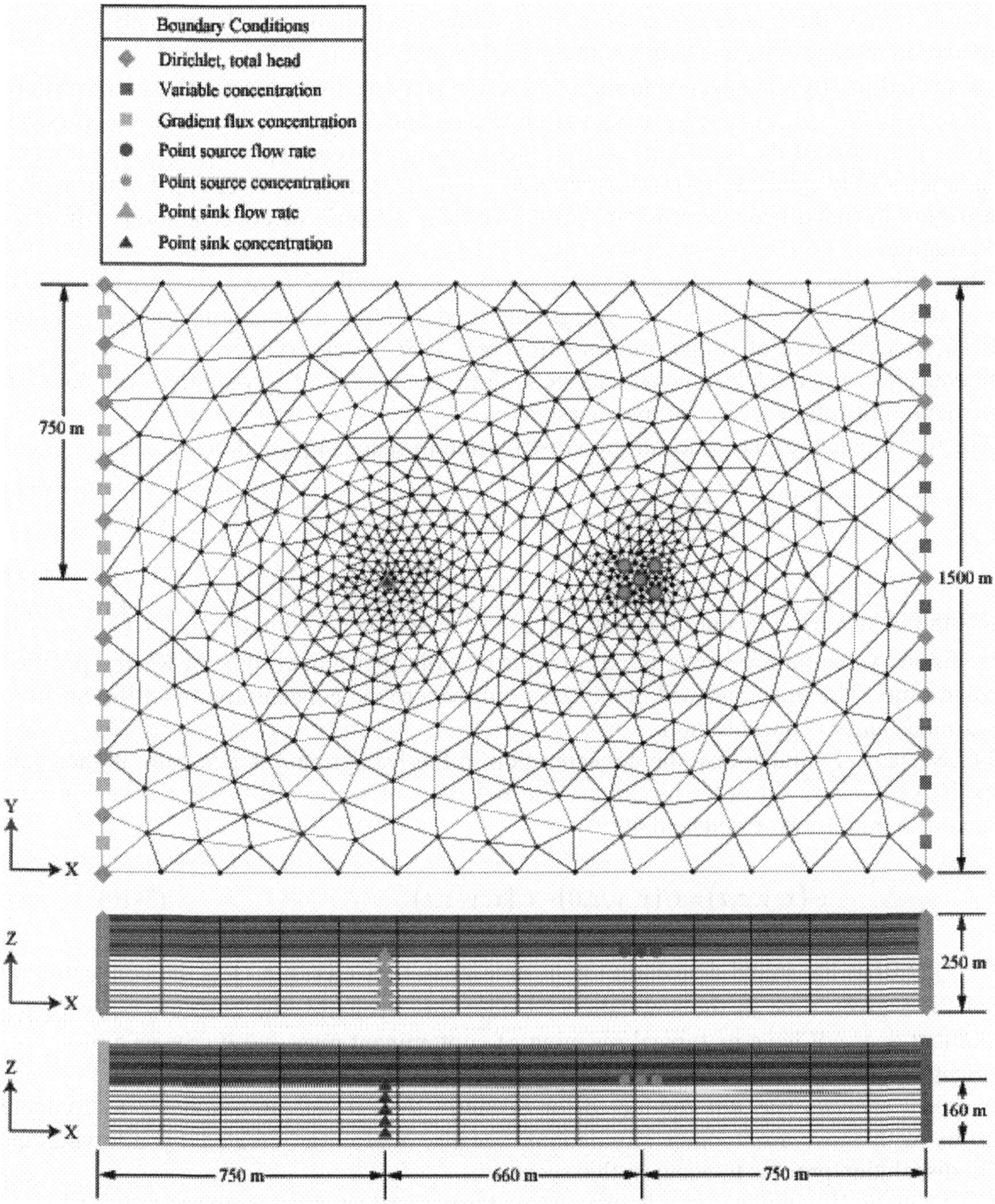

Figure 3.2 Plan and cross-sectional view of the finite element computational domain with placement of initial and boundary condition nodes described in the legend.

Single Domain Model Output

The simulations resulted in breakthrough curves for clean water at the three observation wells. This behavior is contrary to typical contaminant transport modeling exercises in which the breakthrough of increasingly tainted water is observed. Here, it is important to recall that chromium-free water is being injected to an aquifer characterized by a background chromium concentration. Thus, the injection results in an initially clean plume which gradually becomes contaminated on its edges due to mixing with the background water.

Simulated groundwater concentrations were observed at three locations, approximately 200 m, 650 m and 1,100 m down-gradient from the recharge zone. Concentrations at nodes spanning the thickness of the aquifer at each of these locations were depth-averaged to simulate sampling from a fully penetrating production well. However, there is a caveat for the model characterization intended here: information on the vertical behavior of concentration profiles is lost by averaging.

The bulk behavior of recharged water plumes is described here by means of spatial moment analysis. This technique allows integrating model output information at node locations into a large scale measure of solute motion (Freyberg, 1986). Observations of the zeroth and first moments allow the modeler to characterize both the total mass of a contaminant in solution and movement of the plume center of mass, respectively.

The ijkth moment of the concentration distribution in space, M_{ijk}, is defined as

$$M_{ijk}(t) = \iiint_{zyx} \theta \ \hat{C}(x,y,z,t) \ x^i \ y^j \ z^k \ dx \ dy \ dz \qquad (3\text{-}9)$$

where θ is the volumetric moisture content; x y z are the spatial coordinates; and $\hat{C}(x,y,z,t)$ is the concentration value (above background) in units of moles per unit volume of water, specified in space and time. The integrand will be nonzero only where the concentration is different from the background (initial) concentration.

For Scenario I simulations, the concentration in the background is generally higher than that of recharged water. In order to capture the behavior of recharged water the background corrected concentration was calculated as:

$$\hat{C}(x,y,z,t) = C(x,y,z,0) - C(x,y,z,t) \qquad (3\text{-}10)$$

where $C(x,y,z,0)$ is the initial concentration at time zero; and $C(x,y,z,t)$ is the concentration calculated for a given time in the simulation. Under this Scenario, the absolute value of the zeroth moment does not have its typical meaning of contaminant mass but works as a proxy for mass of clean water introduced into the model domain.

Moment analysis was not conducted for Scenario II as it was unreasonable to compare these results with those from other simulations because mass was created by allowing the chromium dissolution process to occur in this scenario.

For Scenario III simulations, where recharged water collected contaminant as it infiltrated against a clean background, corrected concentration was calculated as:

$$\hat{C}(x,y,z,t) = C(x,y,z,t) - C(x,y,z,0) \qquad (3\text{-}11)$$

The first moment, normalized by the mass in solution, defines the coordinate location of the center of mass (x_c, y_c, z_c) of the recharged water plume and was calculated for both Scenarios I and III.

$$x_c = \frac{M_{100}}{M_{000}} \quad y_c = \frac{M_{010}}{M_{000}} \quad z_c = \frac{M_{001}}{M_{000}} \tag{3-12}$$

Preliminary Simulations: Model Sensitivity to Discretization

Spatial Discretization

Four preliminary simulations were run using initial, boundary, and point source/sink boundary conditions of the base case but varying the resolution of the spatial discretization in both the vertical and horizontal dimensions. In the vertical dimension, a *fine* descriptor implies a discretization constructed with 10 layers in the saturated zone and 20 layers in the unsaturated zone. A *coarse* descriptor for the vertical dimension implies 5 layers in the saturated zone and 10 layers in the unsaturated zone. For the horizontal dimension, a descriptor of *fine* implies that a 2D mesh of 2,025 nodes was used to construct the 3D model. A constant time step of 50 days was used for all four simulations.

The impact of spatial discretization on the zeroth moment over time is given in Figure 3.3. During the first 1,800 days of the simulation the amount of clean water in the model domain increases at a constant rate, this is reflected in a linear increase in the magnitude of the zeroth moment. After the recharge period, the zeroth moment is expected to remain constant for as long as the clean water plume remains in the system. However, all throughout this work, simulations show a slight increase in M_{000} that could possibly be attributed to numerical dispersion.

For the space configurations investigated here, resolution in the vertical direction affects the zeroth moment more than horizontal resolution: at both simulations with fine vertical discretization the value of $\boldsymbol{M_{000}}$ at the end of the recharge period is higher than for both simulations with coarse vertical discretization. The finer the discretization (both vertical and horizontal) the higher the resulting value of $\boldsymbol{M_{000}}$ (Figure 3.3) and lower minimum concentration (Figure 3.4). Table 3.2 shows FEMWATER computational times.

Although the finest of these discretizations is expected to give the most accurate results, computational times were considered too long for purposes of the sensitivity analysis. Both intermediate discretizations took similar times but based on the results of the zeroth moment analysis, the fine vertical/coarse horizontal option was deemed most appropriate and was selected as the model domain discretization for the remainder of this work.

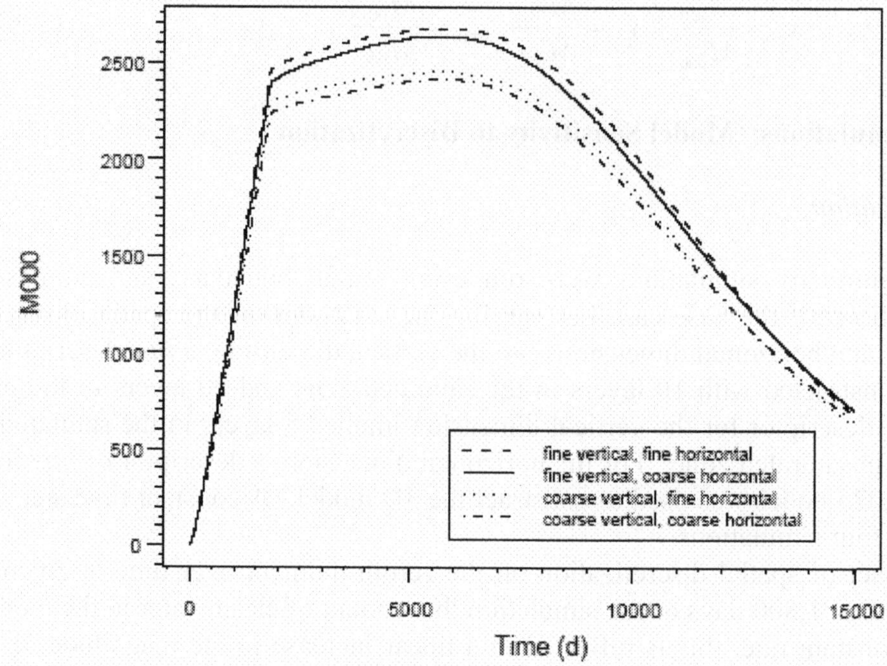

Figure 3.3 The impact of spatial discretization on the zeroth moment (M_{000}) of the clean water (reverse plume) distribution.

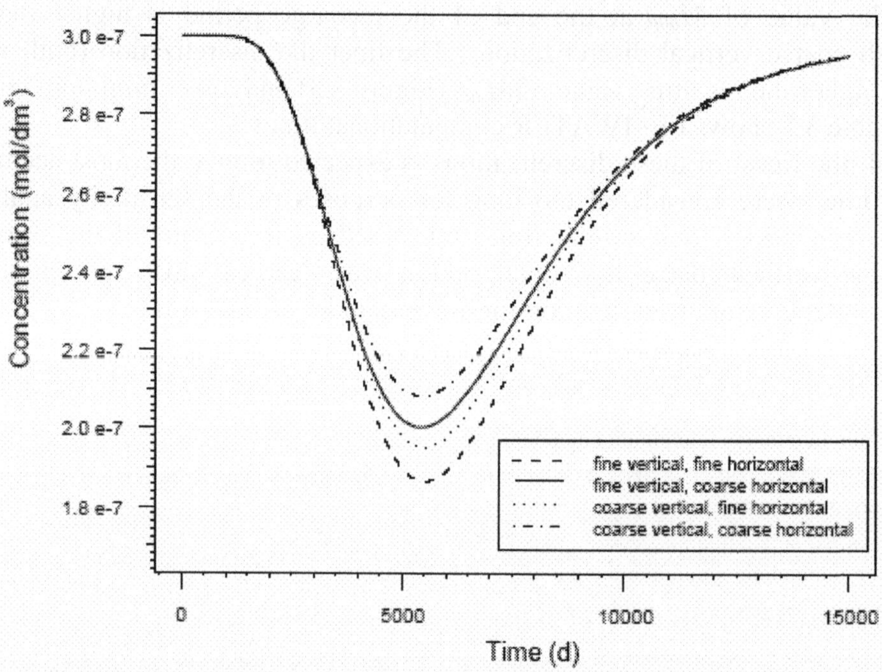

Figure 3.4 The impact of space resolution on concentration at observation location 650 m down-gradient from the recharge zone.

Table 3.2
Computational times for different spatial discretizations

Vertical Dimension	Horizontal Dimension	Total Number of Nodes	Number of Elements	Flow Run Time (min)	Transport Run Time (min)
fine	fine	62,775	117,780	29.5	175.7
fine	coarse	20,615	38,400	6.4	27.3
coarse	fine	32,400	58,890	10.1	27.5
coarse	coarse	10,640	19,200	3.2	8.7

Temporal Discretization

A series of preliminary simulations were completed using initial, boundary and point source/sink boundary conditions of the base case but varying the resolution of the temporal discretization. A constant time step, Δt, was used throughout each simulation, with values ranging between 50 and 500 days. Finer time resolution results in higher values for M_{000} (Figure 3.5). While the zeroth moment is affected by Δt in the range 50-300 days, there is a clear jump in the M_{000} value when Δt is increased to 400 and 500 days. This can be the result of both less accurate model solution and also the fact that the recharge period did not coincide with the end of a time step, therefore resulting in an effective recharge period shorter than the specified 1800 days.

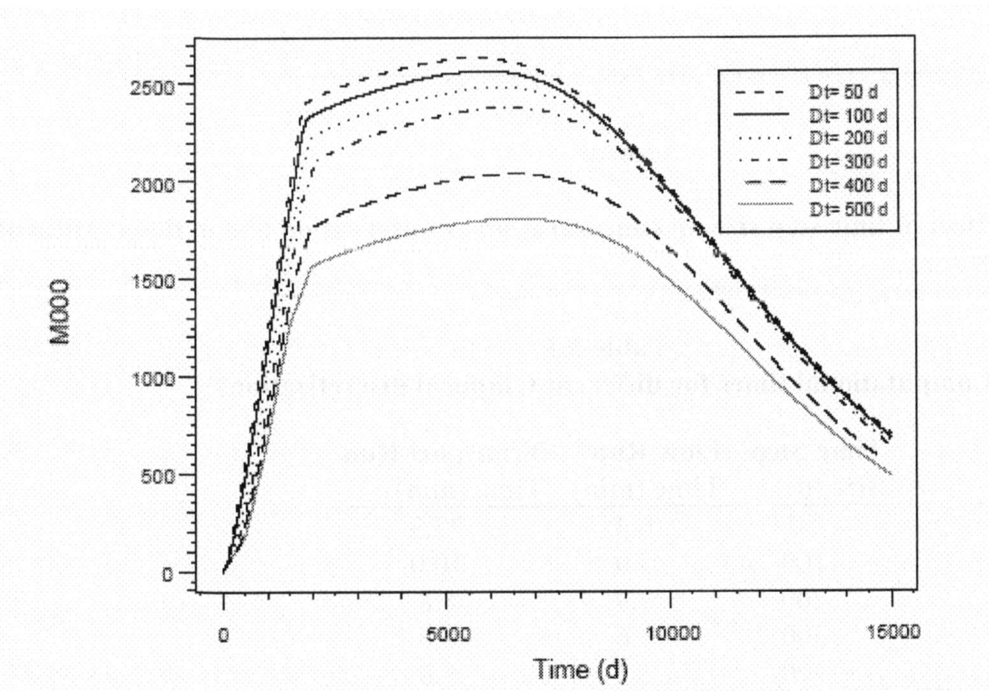

Figure 3.5 The effect of time step size on the zeroth moment calculations.

Finer time resolution also results in lower minimum concentration and shorter time of arrival of the clean water plume at observation location 2 (Figure 3.6). Here again, there is a significant impact of Δt when going from 300 to 400 days. Table 3.3 shows FEMWATER computational times for the six time steps evaluated.

Both the zeroth moment and concentration time series obtained with Δt=100 days closely match the results at Δt=50 days, while the simulation can be completed in almost half the time. For this reason Δt=100 days was selected for the remaining simulations in this work. It is also interesting to note that as the time step size increased, simulations became shorter, except at 500 d, where the smaller number of steps is overcome by longer times necessary to reach convergence.

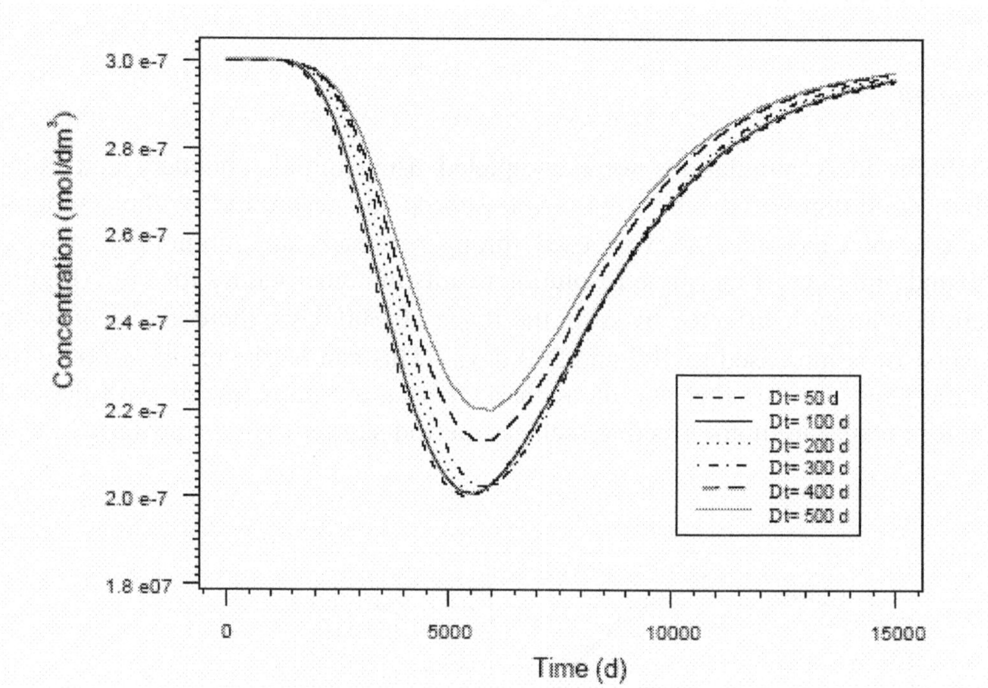

Figure 3.6 The effect of time step size on concentration at observation 650 m downgradient from the recharge zone.

Table 3.3
Computational times for different temporal discretizations

Time Step Size (d)	Flow Run Time (min)	Transport Run Time (min)
50	6.4	27.3
100	5.0	10.0
200	2.9	8.8
300	1.6	6.6
400	1.1	3.0
500	1.6	4.3

Single Domain Parameter Sensitivity Tests

With the single domain grid geometry and spatial and temporal discretization established, the sensitivity of the calculated breakthrough curves and moments to key flow and transport parameters was tested. A range of values for dispersivity, recharge rate and recharge configuration were tested in Scenario I (nonreactive case, i.e., mixing only). For Scenario II (reactive case in the saturated domain), the sensitivity of the output to the zero-order chromium dissolution rate value was investigated for different concentrations values entering the model domain with the background flow: Case A in influent value was set at 3e-7 mol/L and Case B was set at an effectively nondetectable level of 1e-11 mol/L. For Scenario III, the effect of the same chromium dissolution rate parameter was investigated for the unsaturated portion of the single domain.

Single Domain Transport Model Validation

Reactive transport simulations were carried with the code HBGC123D. To analyze the results obtained in the next section in the context of those observed in the previous section for transport in the absence of reaction, the base case simulation for Scenario I was run in HBGC123D and the results were compared to those obtained with FEMWATER.

HBGC123D offers two options for solving the transport equations of aqueous solutes: the conventional Eulerian finite element method and the hybrid Lagrangian-Eulerian (LE) finite element method. In the Lagrangian approach a solute plume is represented by a finite number of moving particles. In the Eulerian approach the advection-dispersion equation is solved over a fixed computational grid. In the Eulerian Lagrangian approach transport is solved in two steps: first the advective portion of the transport equation is solved by a Lagrangian approach, then transport due to dispersion is solved using the Eulerian approach. In some cases, hybrid Lagrangian-Eulerian methods result in more accurate solutions than either Lagrangian or Eulerian algorithms (Ruan and McLaughlin, 1999). Results obtained with both numerical solution techniques are presented here.

Both moments (Figure 3.7) and concentration (Figure 3.8) match up almost perfectly for simulations run with FEMWATER and those with HBGC when using the Lagrangian –Eulerian solution technique, validating a comparison of reactive transport results obtained with the Lagrangian Eulerian method presented in Scenario II to the non-reactive transport base case in Scenario I. Both solution techniques in HBGC resulted in similar computational times (approximately 15 min) which was somewhat longer that the time needed for the FEMWATER simulation (10 min).

DUAL DOMAIN MODELING (UNSATURATED-SATURATED)

In ASR, water is placed within a spreading basin and infiltrates through the vadose zone. The wetting front progresses downward toward the water table and enters into the aquifer, exhibiting a complex hydrodynamics. The computational requirements for modeling flow and transport under unsaturated conditions are much greater than those required for saturated conditions. The higher computational requirements are attributed to the highly nonlinear form the governing equations take under unsaturated conditions.

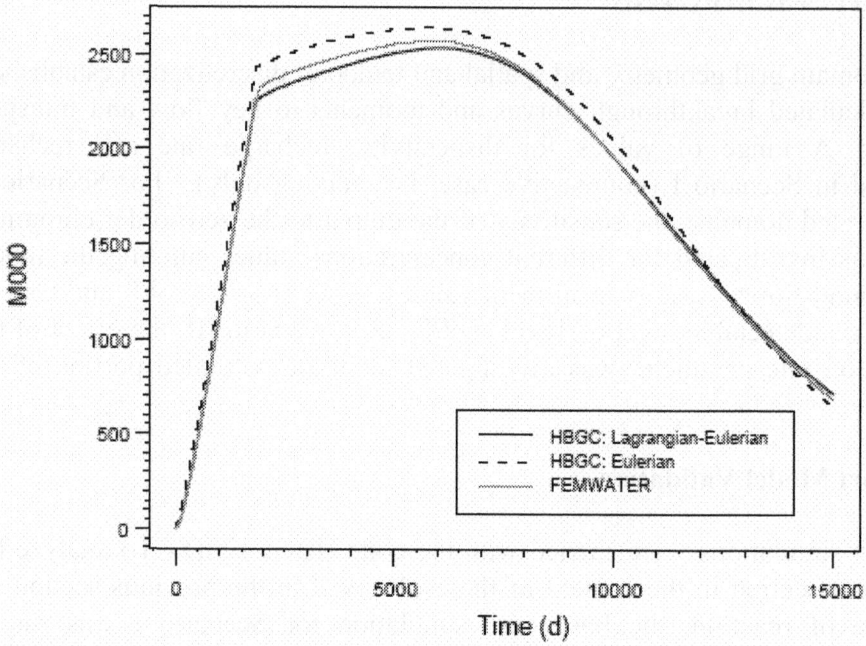

Figure 3.7 The effect of the numerical solution technique on the zeroth moment calculation in reactive transport simulations.

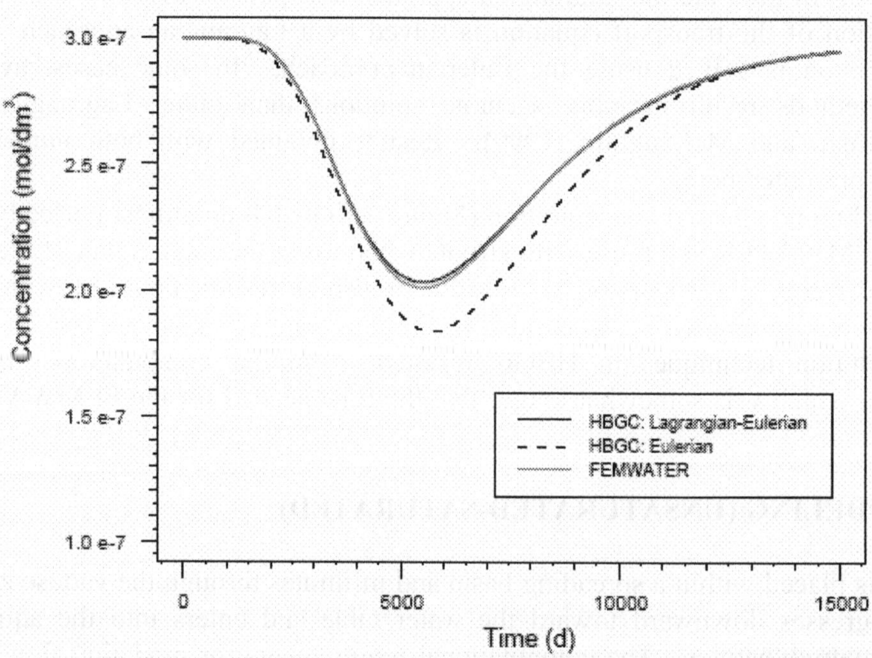

Figure 3.8 The effect of the numerical solution technique on breakthrough concentrations observed 650 m down-gradient from the recharge zone.

These nonlinearities are dealt with by increasing both the spatial and temporal resolutions, thus increasing the computational time required for a simulation. The dual domain modeling approach proposed here is a new approach that decouples the soil domain and groundwater domain simulations, thereby allowing concentration of computational efforts in a smaller soil domain and less intensive simulations in the larger groundwater domain. The key issue then is to develop an approach which executes this plan without sacrificing model accuracy.

The additional cost in computational time, associated with the unsaturated conditions, would be deemed acceptable if the only purpose of the model was to increase the understanding of the hydrologic mechanisms which govern flow and transport within a conjunctive use system. If, however, the model is to also serve as a water resources management tool, placing the model within the framework of an optimization algorithm, the additional computational costs would become unacceptable. For any type of optimization algorithm the number of calls to the simulation model is enormous. A methodology is presented for constructing a model which simulations flow and transport in both the saturated and unsaturated zones while remaining applicable for use within a management optimization algorithm.

The conjunctive use system is separated into two separate models (System 1 and System 2). The schematic for Systems 1 and 2 are shown in Figure 3.9. System 1 models the unsaturated zone from the spreading basin (350 m above datum) to the aquifers water table (183 m above datum). System 2 models the unconfined aquifer from bedrock (0 m above datum) to approximately 7 m above the water table (190 m above datum) with the bulk of the unsaturated zone eliminated from the models spatial domain. System 1 is independent of System 2, however, the recharge boundary of System 2 (190 m above datum) is constructed from a System 1 simulation.

Model parameters which are common in both System 1 and System 2 are shown in Table 3.4. All simulations employ a homogeneous, isotropic medium. The systems saturated permeability was obtained from pump tests conducted during the pilot field study. Soil properties specified for the unsaturated zone are listed in FEMWATER as average values for sandy clay loam.

Table 3.4
Parameter descriptions and values for flow and transport simulations

Description	Symbol	Value	Units
density of water	ρ	997.075	kg/m_3
dynamic viscosity of water	μ	72.962	kg/m/day
compressibility of water	α	0.0	$m \cdot day^2/kg$
saturated permeability	k_s	3.0E-11	m
tortuosity	t	0.7	-
longitudinal dispersivity	a_L	50.0	m
lateral dispersivity	a_T	5.0	m
molecular diffusion coefficient	a_M	0.0000864	m^2/day
soil-specific coefficient	α	0.059	cm^{-1}
soil-specific exponents	β	1.48	-
residual moisture content	θ_r	0.1	-
saturation moisture content	θ_s	0.39	-

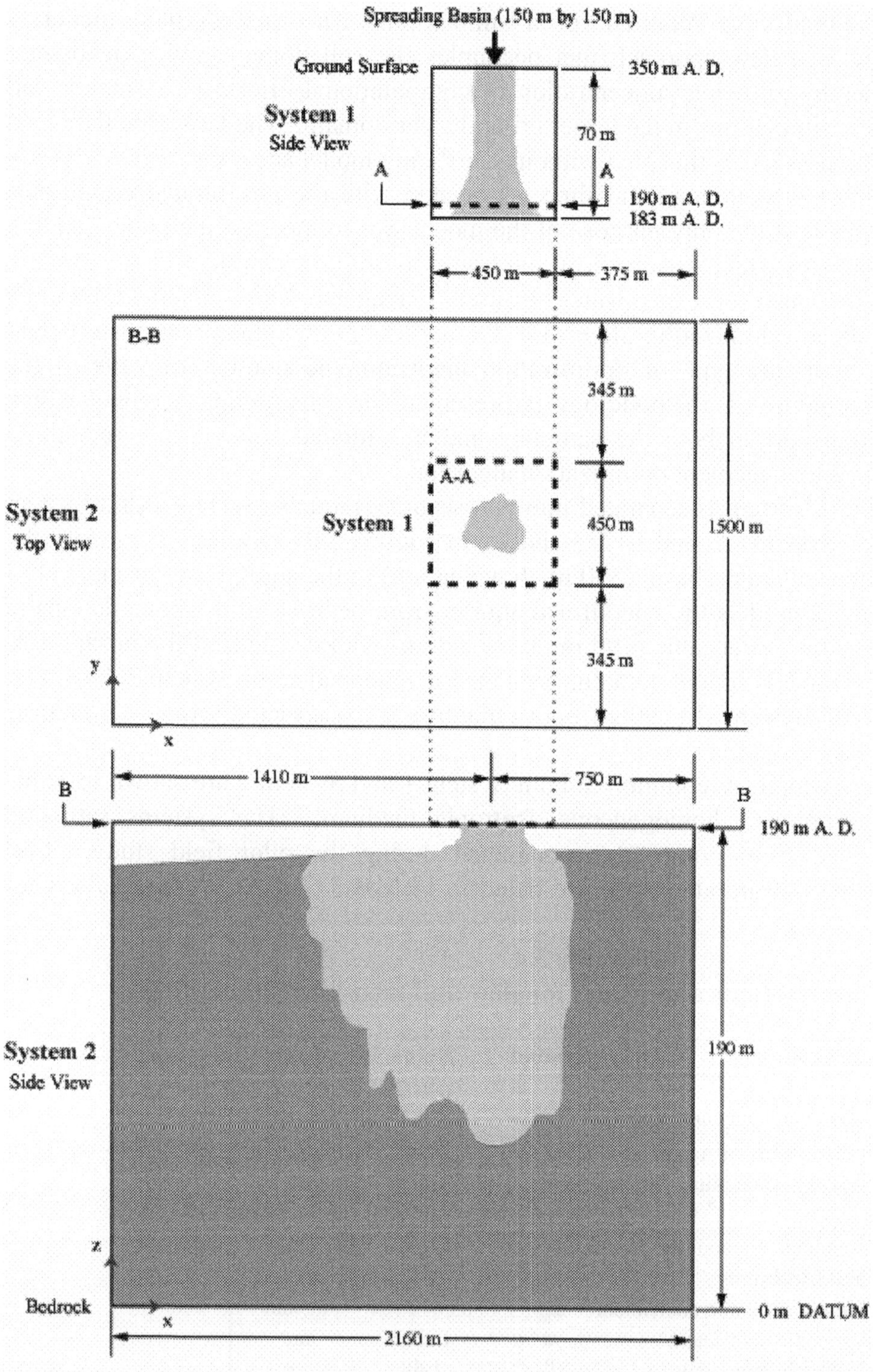

Figure 3.9 Schematic of systems 1 (unsaturated domain) and 2 (saturated domain with 10 times exaggeration in the vertical direction.

Numerical Discretization for Dual Domain Simulations

Temporal Discretizations

The total simulation time for System 1 is 50 days. The temporal discretization of System 1 is shown in Figure 3.10 and consists of five thousand 0.01 day time steps. Time step selection was based on preliminary sensitivity results for flow and transport simulations within the unsaturated zone.

The total simulation time for System 2 is 5000 days. The temporal discretization for System 2 is shown in Figure 3.10 and consists of one-hundred 0.05 day time steps during the first 50 days of simulation and nine hundred and ninety nine 50 day time steps for the duration of the simulation. The use of a shorter time step during the first 50 days of simulation is to (a) better capture the temporal resolution of System 1 within the recharge boundary of System 2, and (b) more accurately describe the flow and transport of recharged water within the unsaturated zone. After 50 days of simulation, the recharged water is primarily within the saturated zone, thus allowing for longer time steps.

Spatial Discretizations

The System 1 model domain is a rectangular box 450 m long, 450 m wide, and 70 m deep (Figure 3.11). The spatial discretization within the model domain consists of 26,532 computational nodes forming 48,860 prismatic elements. There are thirty five 2 m element layers in the box. Any horizontal cross-section of the box contains 1,396 two-dimensional (2-D) triangular elements. The density of 2-d elements is highest within the spreading basin.

The System 2 model domain is a rectangular box 2,160 m long, 1,500 m wide, and 190 m deep (Figure 3.11). The spatial discretization within the model domain consists of 10,557 computational nodes forming 19096 prismatic elements. There are ten 1 m element layers in the upper portion of the box and twelve 15 m thick element layers in the lower portion of the box. Any horizontal cross-section of the box contains 868 2-D triangular elements. The density of 2-D elements is highest within the spreading basin and the area surrounding a potential well site located at X = 750 m and Y = 750 m.

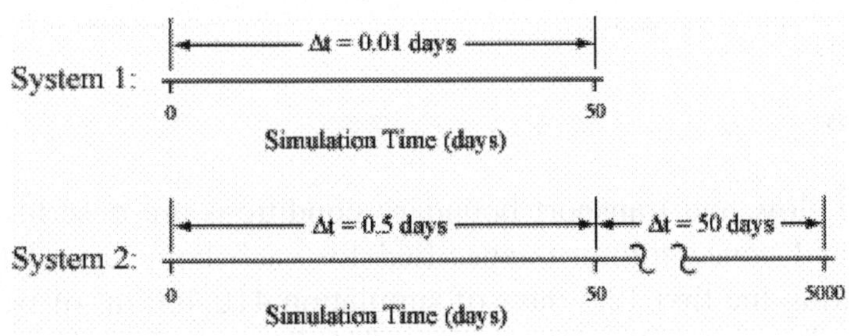

Figure 3.10 Temporal discretizations for Systems 1 and 2.

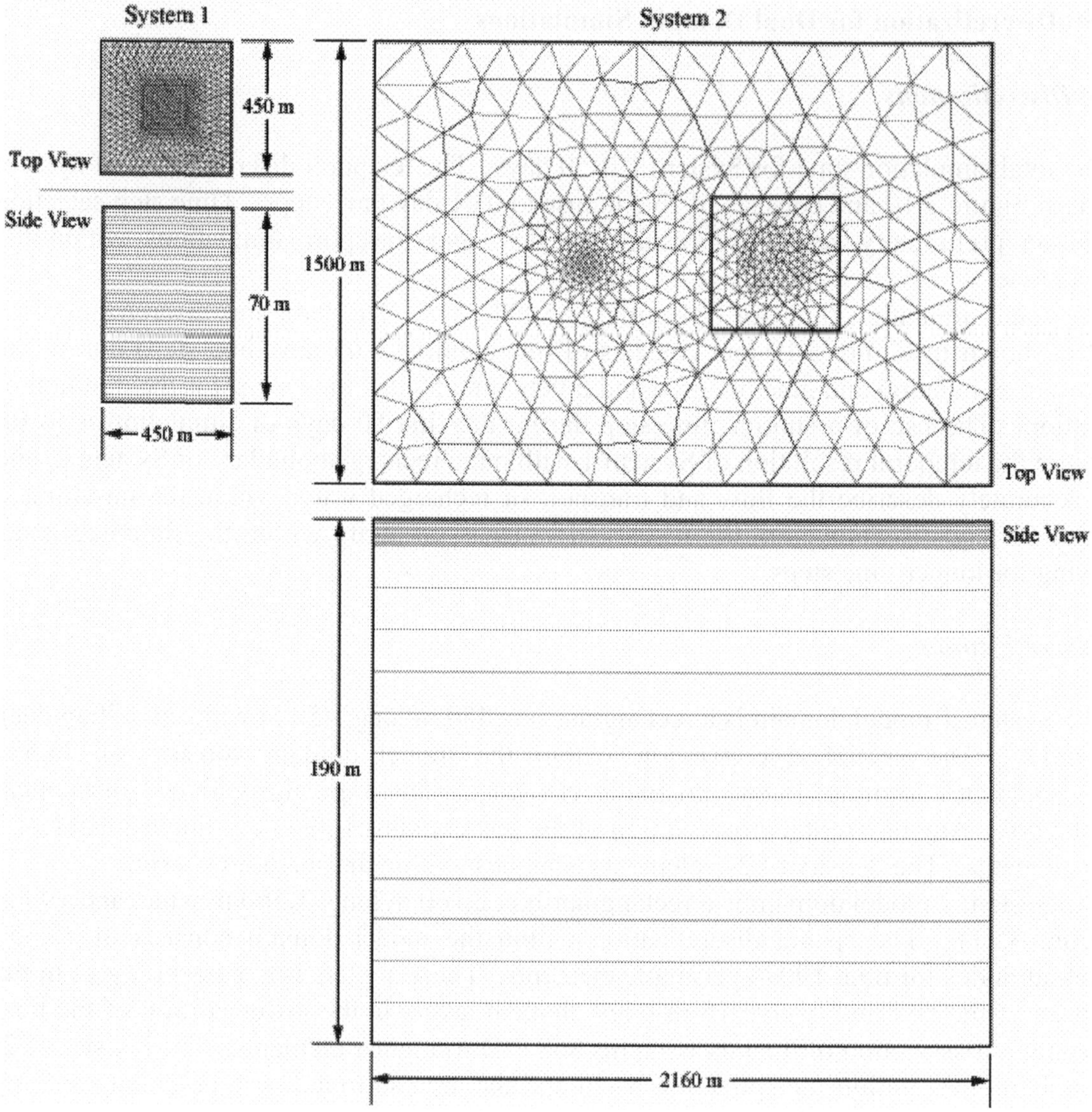

Figure 3.11 Finite element spatial discretizations for Systems 1 and 2.

Boundary Conditions

The system 1 flow and transport boundary conditions are described in Figure 3.12. A transient fluid flux and concentration is specified for each boundary element face within the spreading basin. During the first 12.5 days of simulation 410,400 m^3 of water enters the system through the spreading basin at a Cr concentration of 3E-7 mol/l. The water table is established at the systems base with a total pressure head of 183 m specified for each node within the lower boundary (Z = 183 m). A constant concentration flux gradient of 0 is set for each element face within the lower boundary to prevent the buildup of Cr over time.

The system 2 flow and transport boundary conditions are described in Figure 3.12. On the right and left sides of the system domain constant head boundary conditions were assumed based on groundwater elevation contours measured in the study area. Nodal values for total pressure head are set at 184.4 m and 180.7 m on the right (x = 2,160 m) and left (x = 0 m) boundaries respectively. A constant concentration of 1E-11 is set for each element face within the right boundary and a constant concentration flux gradient of 0 for each element face within the left boundary. Recharge boundary conditions for System 2 are established using the Recharge Boundary Construction Algorithm (see next section).

Initial Conditions

The System 1 initial conditions describe an unsaturated system with a water table established at the base of the model domain. The total pressure head throughout the system is set constant at 183 m. Concentrations within the system are assumed constant at 1E-11 mol/L.

The System 2 initial conditions describe the aquifers baseline flow with an inclined water table. The initial pressure heads are determined from a steady state solution of the model with no recharge. Concentrations within the system are assumed constant at 1E-11 mol/L.

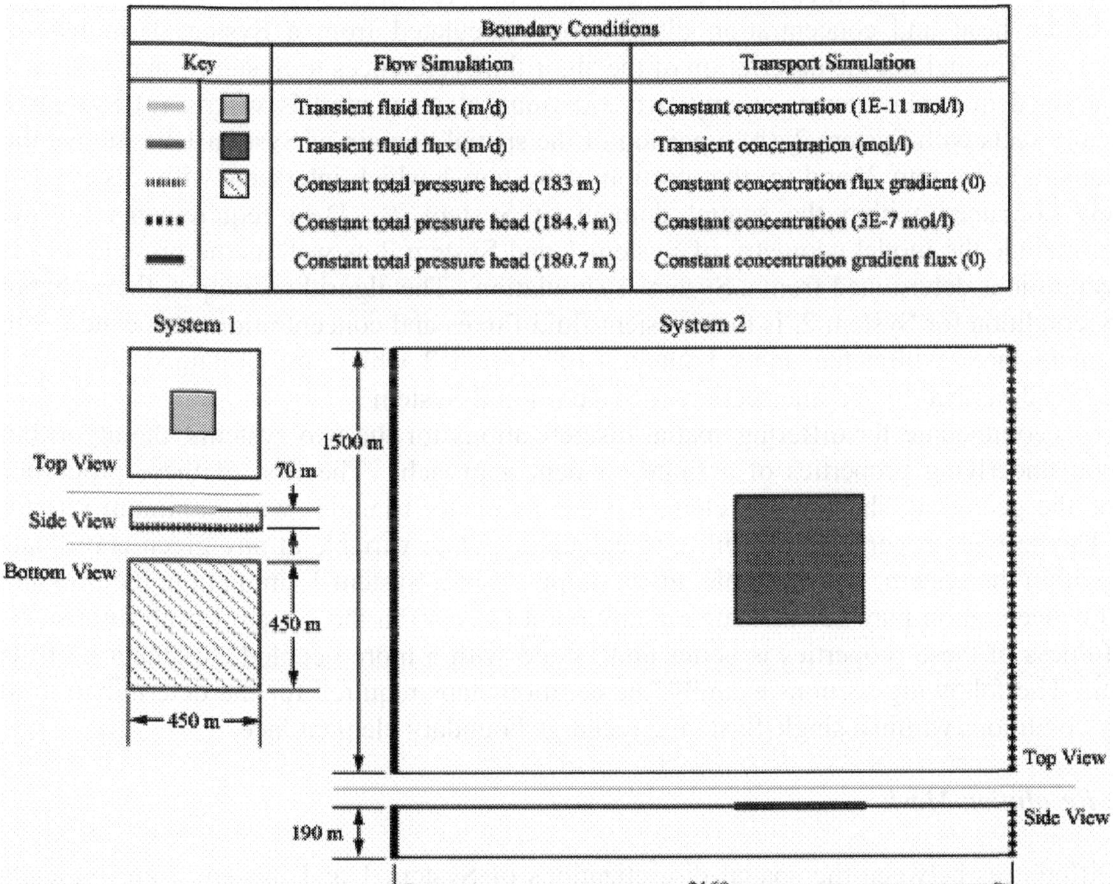

Figure 3.12 Boundary conditions for Systems 1 and 2, no exaggeration in the z-dimension.

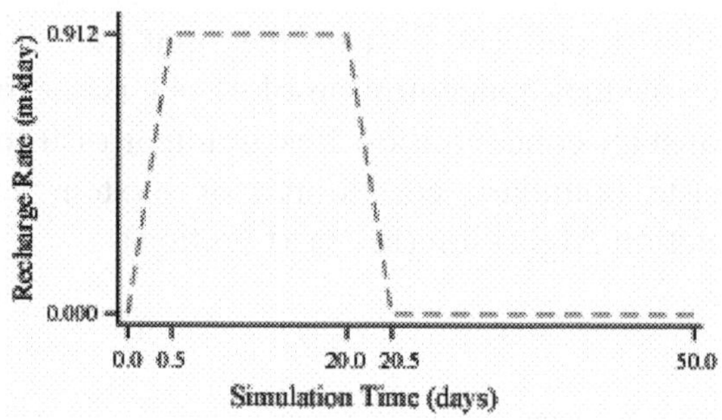

Figure 3.13 Transient recharge boundary condition for System 1.

Recharge Boundary Construction

An algorithm is presented for the construction of a recharge boundary within System 2 using pressure head and concentration distributions calculated from a System 1 simulation (Figure 3.13). Throughout the description of the algorithm, references to System 1 and System 2 may be defined in general terms as any two 3-d finite element models where, (a) System 1 spatially intersects with System 2, (b) a portion of the spatial domain of System 1 lies above the spatial domain of System 2, and (c) that portion of System 1 which intersects with System 2 is completely contained within the spatial domain of System 2. Data requirements for the algorithm include the model geometry of System 1 and System 2 as well as the pressure heads and concentrations determined from a System 1 simulation. The algorithms output, the recharge boundary condition for System 2, is the transient fluid fluxes and concentrations associated with those element faces within the upper boundary of System 2 which are contained within the spatial region described by the intersection of System 1 and System 2.

To accommodate for differing spatial discretizations for the two systems, the algorithm relies upon underlying properties of a finite element approach. The first of these properties constrains the geometric shape of all element faces as planer triangles with minimum interior angles. The second property is the ability to calculate a state variable at any given coordinate location within the system. For example, given output from a System 1 simulation, the pressure head and concentration may be calculated at any point (x, y, z) in the system for all time steps. The usefulness of these properties is better understood with a more detailed explanation of the algorithm. The following sections examine the solution steps required for the determination of boundary conditions within a single System 2 recharge boundary element face.

Multiple Resolution Mesh

Differences between the spatial discretizations of System 1 and System 2 are typically due to a much higher spatial resolution within System 1. A search technique is used to capture the higher resolution of System 1 within the System 2 recharge boundary. The first requirement for implementing the search technique is the generation of a sequence of 2-d finite element meshes, where every mesh within the sequence shares a common spatial domain defined by the recharge boundary element face. Mesh placement within the sequence is a function of the meshes spatial resolution. This functional relationship between the level of resolution, i, and the total number of elements, e_i, within a mesh is expressed as

$$e_i = 4^{i-1}, \quad i = 1,\ldots,i_{max} \tag{3-13}$$

where i_{max} is a user defined maximum level of resolution. For example, setting $i_{max} = 3$ results in three separate meshes corresponding to three levels of resolution. The first of these meshes, $i = 1$, contains a single element, the dimensions of which are identical to the original recharge element face of System 2.

Mesh construction for levels of resolution greater than one use as their template the mesh constructed from the previous level of resolution. Figure 3.14 shows the node and element indexing for a sequence of meshes corresponding to three levels of resolution. What ties the three meshes together is the indexing for nodes of identical spatial placement. For example, the spatial locations of nodes 1, 2, and 3 remain constant for all three meshes.

The methodology for mesh refinement going from i to i + 1 is as follows: (a) Select the first element in mesh i. (b) Calculate the midpoints for line segments formed when connecting the element vertices, P. Where, M_1, M_2, and M_3 are the respective midpoints for line segments connecting P_2 to P_3, P_3 to P_1, and P_1 to P_2. (c) Construct new elements in mesh i + 1 defined by

$$\begin{aligned} E_1 &\equiv P_1, M_3, M_2 \\ E_2 &\equiv M_3, P_2, M_1 \\ E_3 &\equiv M_2, M_1, P_3 \\ E_4 &\equiv M_1, M_2, M_3 \end{aligned} \tag{3-14}$$

(d) Identify the element numbers for the newly created elements. Recall that there are four elements produced from the refinement of a single element. (e) Step through each of the newly formed elements assigning node numbers to those points not already identified as nodes. (f) Store the element and nodal information for later use and repeat the procedure for every element in the mesh.

Flow and Transport Calculations

Calculation of the flow and transport through a single System 2 recharge boundary element face requires both a multiple resolution mesh (Section 3.1) and the ability to determine the pressure head and concentration at any point within System 1. The mathematical formulation for the calculation of a state variable within the domain of the system is excluded from this report do to the reliance of the formulation on the specific type of basis functions used within the 3-d finite element model. The general concept behind the formulation is, however, presented in most introductory finite element books.

The ability to determine h and C at (x, y, z) allows for the calculation of fluid flux through any 2-d element within the System 1 spatial domain (Figure 3.15). Darcy's Law, an empirical relationship for flow in porous media, gives the fluid flux, q, as

$$q_c = -\left(\frac{\rho g}{\mu}\right)\mathbf{k}_s k_r A_e \left(\frac{h_{2,c} - h_{1,c}}{Z_2 - Z_1} + 1\right) \quad \text{for } c = 1,\ldots,T_1 \tag{3-15}$$

where c is the time step index for System 1, T_1 is the total number of time steps in the System 1 simulation, A_e is the elements area, and h_1 and h_2 are the hydraulic heads at a distance of ΔZ

above and below the centroid of the element, respectively. The magnitude of ΔZ is specified by the user. It is important to note that Equation 3.15 only describes vertical flow through the element with horizontal flow assumed negligible. The equation could, however, be easily reformulated to accommodate for flow in multiple directions.

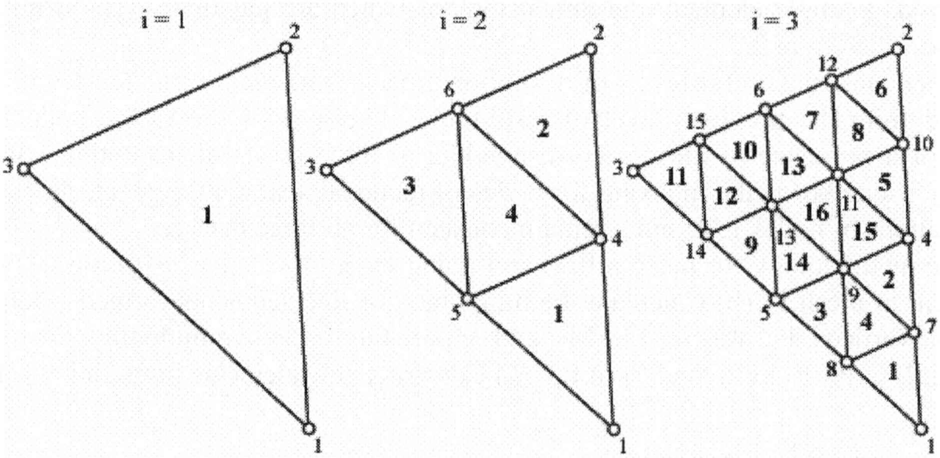

Figure 3.14 Node and element indexing for mesh generation.

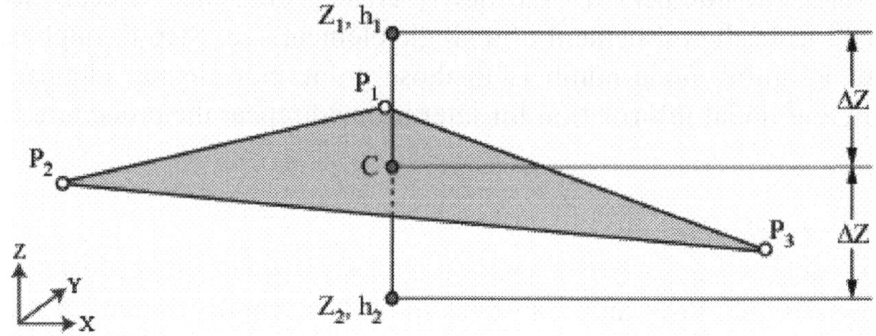

Figure 3.15 Pressure head gradient calculation.

The fluid mass passing through the element during a single System 1 time step is defined as

$$m_c = q_c \rho \left(t_{1,c} - t_{1,c-1} \right) \quad \text{for } c = 1,\ldots,T_1 \qquad (3\text{-}16)$$

where $t_{1,c-1}$ and $t_{1,c}$ are the respective start and end times for time step c. Summing m_c over all time steps in the simulation gives the total liquid mass, M, expressed as

$$M = \sum_{c=1}^{T_1} m_c \tag{3-17}$$

The constituent mass, $\tilde{m}_c$, passing through the element during a single System 1 time step is defined as

$$\tilde{m}_c = C A_e q_c \left(t_{1,c} - t_{1,c-1} \right) \quad \text{for } c = 1,\ldots,T_1 \tag{3-18}$$

where C is the concentration at the centroid of the element at the end of time step c.

Mass values are then initialized as multi-dimensional arrays linking m_c, M, and $\tilde{m}_c$ to the multiple resolution mesh. The mass arrays are expressed as $m_{c,i,j,k}$, $M_{i,j,k}$, and $\tilde{m}_{c,i,j,k}$ where i is the current resolution level, j is the element index for the current resolution level, and k is an array of stored element indices for all resolution levels. Figure 3.16 gives the M indexing (i, j, k) within three resolution levels. The translation between array indexes and element numbers, e, is defined as

$$\begin{aligned} &\hat{e}_1 = 1 \\ &\hat{e}_f = 4(\hat{e}_{f-1}) - 4 + k_f \quad \text{for } f = 2,\ldots,i \\ &e = \hat{e}_i \end{aligned} \tag{3-19}$$

where $\hat{e}$ is a temporary single array.

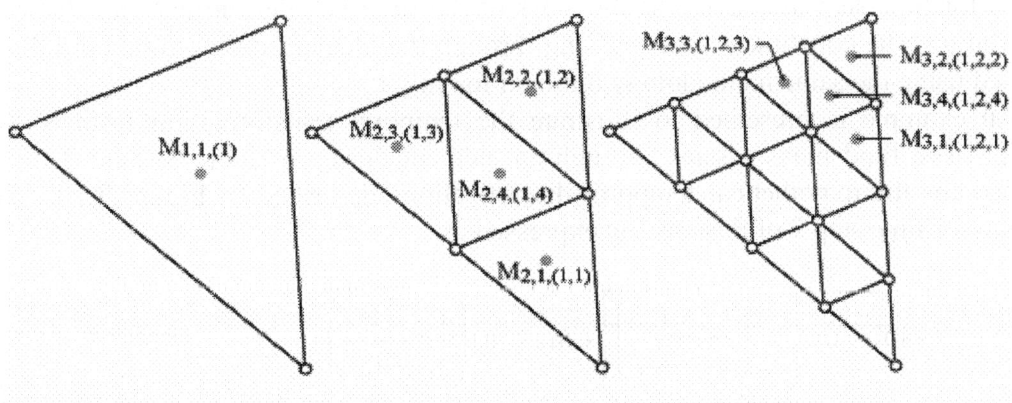

Figure 3.16 Total fluid mass indexing for three levels of resolution.

Search Algorithm

A search algorithm, based on a mass balance approach, is used to calculate the total mass moving through a single System 2 recharge boundary element face. A flow chart for the algorithm is given in Figure 3.17. The premise for the algorithm is the collection of state variable information within System 1 at a level of resolution controlled by the user. The user defined input parameters are: (a) Maximum Percent Difference, TOL, a tolerance level for the total mass percent difference between two resolution levels. Where percent difference, P, is calculated for every iteration of the mesh refinement process and expressed as

$$P = \left[\frac{\left| M_{i,j,k_i} - M_{sum} \right|}{0.5\left(M_{i,j,k_i} + M_{sum} \right)} \right] 100 \tag{3-20}$$

where

$$M_{sum} = \sum_{j=1}^{4} M_{i+1,j,k_{i+1}} \tag{3-21}$$

If the percent difference is less than TOL, the search moves to a higher resolution level, otherwise, the refinement process halts. (b) Maximum Resolution Level, i_{max}, an upper limit on the number of possible mesh refinements. (c) Lower Mass Limit, M_{low}, a lower limit on M_{sum}. If M_{sum} is less than M_{low} the refinement process halts.

Temporal Transformation

At completion of the search algorithm the temporal differences between System 1 and System 2 are addressed. Mass values are summed over the System 2 time steps assuming a finer temporal discretization within System 1. A flow chart for the temporal transformation between System 1 and System 2 is given in Figure 3.18 where d is the time step index for System 2, T_2 is the total number of time steps in the System 2 simulation, $t_{2,d-1}$ and $t_{2,d}$ are the respective start and end times for time step d, and m'_d is the mass of the fluid passing through the element during time period d. The constituent mass passing through the element during time period d, $\tilde{m}'_d$, is determined using the same methodology outlined for m'_d.

An example is presented to illustrate the temporal transformation from System 1 to System 2. The fluid mass passing through an element during a single System 2 time step is determined for the hypothetical temporal discretizations described in Figure 3.19. The mass calculation for time step 2 of System 2 is expressed as

$$m'_2 = m_3 \left[1 - \left(\frac{t_{2,1} - t_{1,2}}{t_{1,3} - t_{1,2}} \right) \right] + m_4 + m_5 \left(\frac{t_{2,2} - t_{1,4}}{t_{1,5} - t_{1,4}} \right) \tag{3-22}$$

After the temporal transformation the fluid mass, m'_d, is converted back to a fluid flux, q, using

$$q_d = \frac{m'_d}{\rho \left(t_{2,d} - t_{2,d-1} \right) A_{elem}} \quad \text{for } d = 1, \ldots, T_2 \tag{3-23}$$

where A_{elem} is the area of the System 2 recharge boundary element face. In addition, the constituent mass, $\tilde{m}'_d$, is converted to a concentration, C, using

$$C_d = \frac{\tilde{m}'_d}{q_d \left(t_{2,d} - t_{2,d-1} \right) A_{elem}} \quad \text{for } d = 1, \ldots, T_2 \tag{3-24}$$

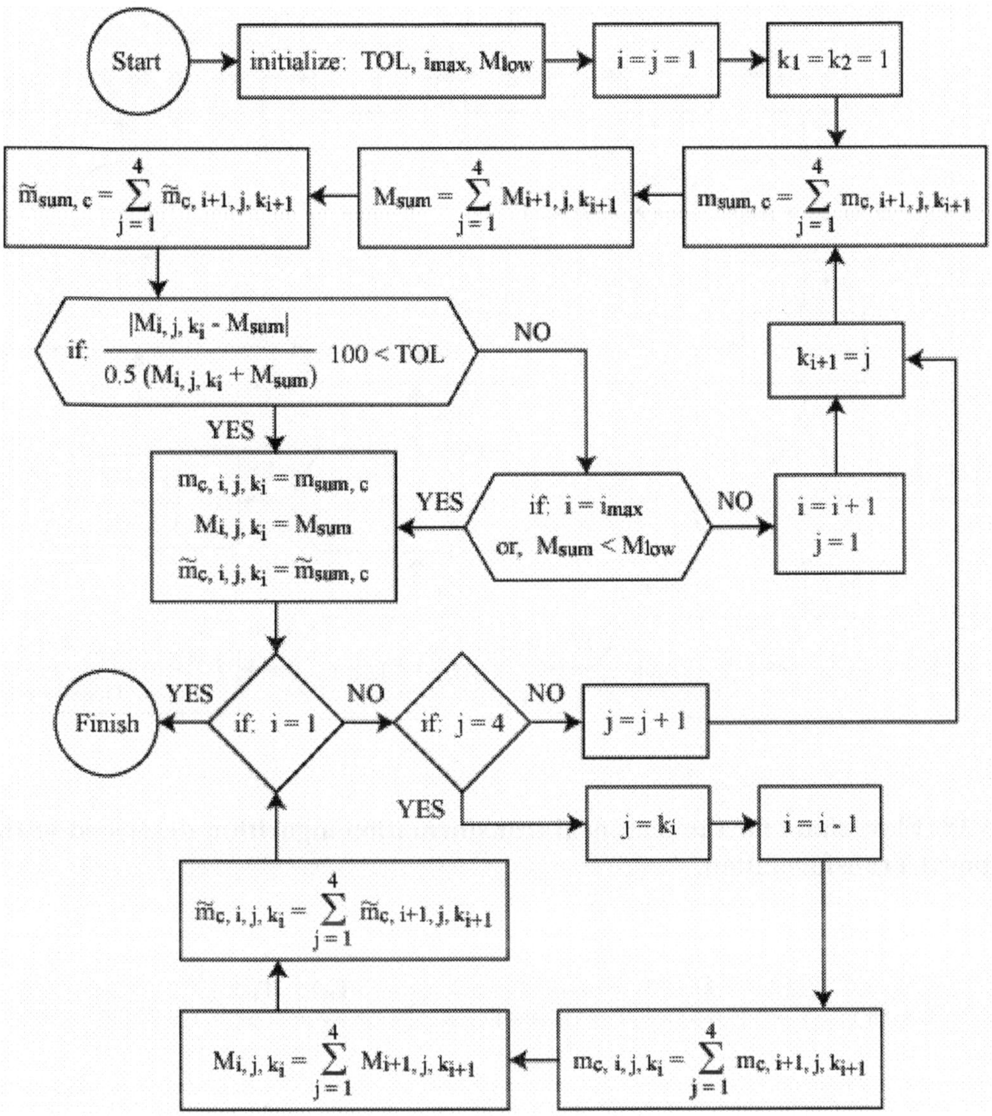

Figure 3.17 Flow chart for the search algorithm described in section on Multiple Resolution Mesh.

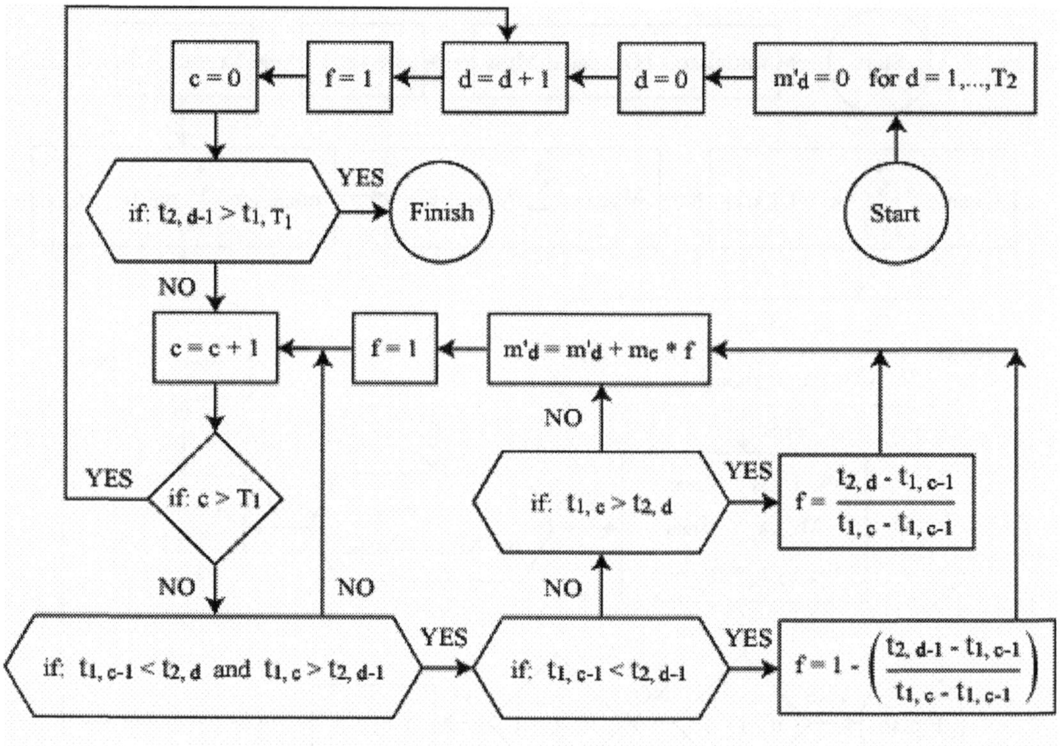

Figure 3.18 Flow chart for the temporal transformation algorithm described in the section on Temporal Transformation.

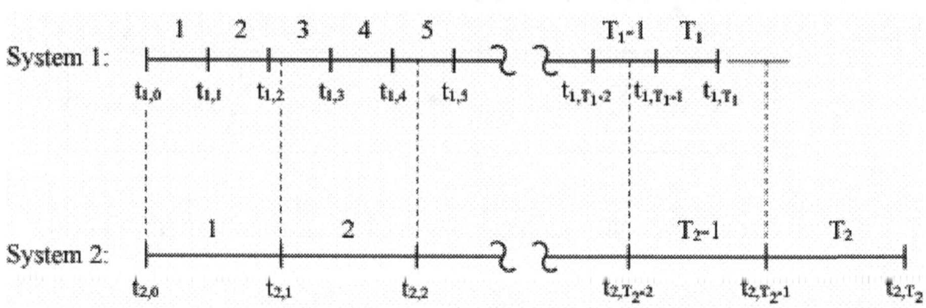

Figure 3.19 Example temporal discretizations for System 1 and System 2.

SUMMARY

This chapter describes two numerical approaches for investigating the potential impact of natural chromium sources on aquifer storage and recovery operations (ASR). The first system is modeled using a single domain that encompasses both the unsaturated and saturated zones of the subsurface. A second modeling approach was investigated as a means of reducing the computational burden imposed by the single domain approach, while preserving the numerical accuracy of the simulation. This is a dual domain approach where an output water flux from the unsaturated domain is adapted to become the input for the saturated domain model. A novel numerical technique for transferring the water flux from the unsaturated zone simulation to the saturated zone simulation is presented.

In Chapter 5, parameter sensitivity analysis results for the single domain modeling approach are presented and discussed with respect to the potential impact of natural metal contamination on ASR operations. This is followed by a several simulations aimed at illustrating the potential utility of the latter approach in solving the same problem.

CHAPTER 4
RESULTS AND DISCUSSION FOR EXPERIMENTAL STUDIES

This chapter presents the findings on several types of experimental studies. In a treatability study, the efficiency of Cr(VI) removal from oxygenated waters by ferrous salts was investigated. Studies of chromium speciation included a laboratory study of the filterability of colloidal Cr(III), an assessment of the contribution of a colloidal Cr fraction to Cr measured in field samples, and a preliminary investigation of field methods for separation of Cr(III) and Cr(VI). Extensive laboratory studies were conducted to investigate the release of Cr from Cr(III)-containing solids under oxidative and non-oxidative conditions. Based on some of this work, geochemical modeling was performed to investigate both the general aspects of the evolution of the chemical composition of recharge water and a specific example of the effects of sorption on dissolution reactions.

REDOX-ASSISTED COAGULATION FOR REMOVAL OF Cr(VI) FROM DRINKING WATER

Since the results of this study have been published (Lee and Hering, 2003), only a brief summary is provided here. The reader is referred to the published work for further details.

This study was based on previous observations that Cr(VI) could be stoichiometrically reduced to Cr(III) by Fe(II) in the absence of O_2 (Buerge and Hug, 1997; Fendorf and Li, 1996; Pettine et al., 1998; Schlautman and Han, 2001; Sedlak and Chan, 1997). Cr(VI) removal by Fe(II) in the presence of O_2 had also been demonstrated as a means of treating Cr(VI)-contaminated water but, in these cases, the target Cr concentrations were usually 50 μg/L (Aktor, 1994; Eary and Rai, 1988; El-Shoubary et al., 1998; Philipot et al., 1985; Sorg, 1979; Zotter and Licsko, 1992). The objective of the current study was to determine whether sub-ppb concentrations of dissolved Cr(VI) could be achieved by redox-assisted coagulation with Fe(II). The efficiency of the desired reaction (for pH $\geq$ 6.5)

$$CrO_4^{2-} + 3Fe^{2+} + 8H_2O \rightarrow 4Fe_{0.75}Cr_{0.25}(OH)_3(s) + 4H^+$$

will depend on the extent to which the competing reaction

$$4Fe^{2+} + O_2(aq) + 4H_2O \rightarrow 4Fe(OH)_3(s) + 8H^+$$

consumes the reductant Fe^{2+}. The relative rates of the two competing reactions will be a function of pH and the (initial) concentrations of the reactants. Loss of Fe^{2+} to reaction with O_2 can, however, be overcome by increasing the Fe(II) dose.

The efficiency of Cr(VI) removal by Fe(II) was tested in three types of solutions: (1) a simple electrolyte (0.01 M $NaNO_3$ buffered with 1 mM PIPES), (2) synthetic Colorado River Water (CRW) (0.01 M $NaNO_3$, 2 mM $NaHCO_3$, 2 mM $Ca(NO_3)_2$), and unfiltered natural CRW. In general, removal efficiencies were assessed after a fixed reaction time (30 min stirred, 15 min unstirred). However, reaction kinetics were also assessed under limited conditions. Complementary experiments were conducted to determine the extent of Cr(VI) adsorption onto Fe(III) oxyhydroxides and the filterability of Cr(III) and Cr(III),Fe(III) oxyhydroxide precipitates.

Reduction of Cr(VI) to Cr(III) was found to be critical to effective removal of Cr from solution. At pH 8.2, Cr(VI) (at an initial concentration of 52 μg/L) was poorly removed from synthetic CRW by sorption/co-precipitation with Fe(III) (at up to 5.6 mg/L as Fe). In contrast, Cr(III) could be removed to below the detection limit (0.5 μg/L) by sorption/co-precipitation with even 0.17 mg/L Fe(III). In the absence of Fe, Cr(III) was found to form colloidal precipitates (in supersaturated solutions) as demonstrated by the lower Cr concentrations in ultrafiltrates as compared with 0.2 or 0.02 μm-filtrates. However, colloidal precipitates were not formed in the presence of Fe(III).

The efficiency of Cr(VI) removal by Fe(II) was found to depend on the molar ratio of Fe-to-Cr. A stoichiometric dose of Fe(II) achieved only poor (ca. 30%) removal of Cr(VI) (at an initial concentration of 52 μg/L) from synthetic CRW at pH 8.2. The residual Cr left in solution was found to be present as Cr(VI) thus the inefficiency in Cr(VI) removal was shown to correspond directly to the incomplete reduction of Cr(VI) to Cr(III).

The removal of Cr(VI) from the simple electrolyte by Fe(II) as a function of pH (for $6.5 \leq pH \leq 8.2$) was in reasonable agreement with that predicted using a kinetic model for the reactions of Fe(II) with Cr(VI) and O_2. This confirmed that the efficiency of Cr(VI) removal was controlled by the extent of the side reaction of Fe(II) with O_2 (for pH values between 7 and 8.2). Below pH 7, the slow kinetics of reaction of Fe(II) with (both) Cr(VI) (and O_2) result in poor efficiency of Cr(VI) removal (over a fixed reaction time). However, previously-reported kinetic constants for the reaction of Cr(VI) with Fe(II) are not entirely consistent. Thus, estimations of Cr(VI) removal efficiency based on kinetic models must be viewed with some caution.

Cr(VI) was observed to be less efficiently removed from natural or synthetic CRW than from the simple electrolyte solution. In all cases, the inefficiencies of Cr(VI) removal at low excesses of Fe(II) (i.e., over the stoichiometric dose) could be overcome by increasing the Fe(II) dose. The Fe(II) dose needed to remove Cr(VI) to the detection limit increased in the order: simple electrolyte < synthetic CRW < natural CRW. However, in all cases, an Fe-to-Cr molar ratio of 30 (corresponding to an Fe(II) dose of 1.7 mg/L as Fe) was sufficient to remove Cr(VI) at an initial concentration of 52 μg/L to below the detection limit (corresponding to >99% removal).

CHROMIUM SPECIATION

Most attention on chromium speciation in water samples has focused on the distinction between Cr(III) and Cr(VI). In principle, the occurrence of (truly) dissolved Cr(III) is limited by the solubility of $Cr(OH)_3(s)$ and should not exceed 7.5 μg/L (as discussed in Chapter 1). Thus, higher concentrations of total, dissolved Cr should be attributable either to Cr(VI) or organically-complexed Cr(III), with the latter species being unlikely to occur in groundwaters with low dissolved organic carbon (DOC) (Kaczynski and Kieber, 1994). However, water samples collected for measurements of "dissolved" metals are commonly filtered through 0.45 or 0.2 μm filters, which have been shown to allow the passage of colloidal particles through the filters (Benoit and Rozan, 1999). This possible artifact was assessed for laboratory and field samples.

Filterability Studies with Colloidal Cr(III)

The passage of colloidal Cr(III) through filters of varying pore sizes was investigated using a solution supersaturated with respect to amorphous $Cr(OH)_3(s)$ (51.0 μg/L Cr(III), pH =

8.2, 850 mg/L $NaNO_3$). As can be seen from Table 4.1, only the ultrafilter produced a filtrate in which the measured Cr concentrations were in reasonable agreement with the calculated concentration of dissolved, inorganic Cr(III) in equilibrium with $Cr(OH)_3(s)$. Cr concentrations in the filtrates obtained with 0.2 and 0.02 μm filters were significantly higher, indicating substantial passage of colloidal Cr(III) through these filters.

Table 4.1
Filterability of colloidal Cr(III)

filter pore size	% of Cr(III) passing through filter	[Cr(III)] in filtrate
0.2 μm	59.2	30.2
0.02 μm	49.0	25.0
ultrafilter (0.003 μm nominal)	8.2	4.2
equilibrium concentration of dissolved, inorganic Cr(III)		7.5

Comparison of Filter Performance with Field Samples

The concentrations of Cr measured in filtrates of field samples obtained using 0.2 μm filters and ultrafilters were compared to assess the possible contribution of colloidal Cr to "dissolved" concentrations. Water samples were collected from a series of springs in May 2002 and, by MWDSC personnel, from three wells in January 2002. Measurements of pH were made in the field and samples were filtered and analyzed at Caltech.

Cr concentrations in spring water samples were uniformly low (Table 4.2). All the springs sampled were overgrown with algae and/or higher plants. Reduction of Cr(VI) to Cr(III) by plants and organic matter has been reported (Elovitz and Fish, 1994; Lytle et al., 1998; Wittbrodt and Palmer, 1996) and thus might be expected in these systems. If such reduction had occurred, it would be expected that the Cr(III) would be lost to surfaces by sorption and/or precipitation and thus dissolved Cr concentrations would be low. No significant differences in Cr concentrations were observed between the unfiltered samples and samples filtered through 0.2 μm filters or ultrafilters indicating that the Cr was not present in colloidal form.

Water samples collected from wells contained substantially more Cr than the spring water samples (Table 4.3). The presence of colloidal Cr in these samples could be excluded on the basis of the similarity between Cr concentrations in unfiltered, 0.2 μm-filtered, and ultrafiltered samples.

Table 4.2
Chromium concentrations in spring water

sample site	pH	sample treatment	Cr concentration (μg/L)
Bonanza spring	8.3	unfiltered	1.16
		0.2 μm	1.04
		ultrafiltered	1.01
Rock spring	8.4	unfiltered	0.40
		0.2 μm	0.45
		ultrafiltered	0.62
stock tank	8.4	unfiltered	0.78
		0.2 μm	0.62
		ultrafiltered	0.63
van Winkle spring	7.3	unfiltered	1.04
		0.2 μm	0.99
		ultrafiltered	1.01

Table 4.3
Chromium concentrations in well water samples

sample ID	pH	sample treatment	Cr concentration (μg/L)
well 33	8.02	unfiltered	27.4
		0.2 μm	27.7
		ultrafiltered	28.3
office	7.89	unfiltered	23.3
		0.2 μm	22.7
		ultrafiltered	23.4
well 21-N	8.10	unfiltered	27.4
		0.2 μm	27.2
		ultrafiltered	26.7

Preliminary Studies of a Field Method for Separation of Cr Species

A variety of methods developed for the separation of Cr(III) and Cr(VI) have been reviewed previously (Kotas and Stasicka, 2000). Most of these methods use anion- or cation-exchange columns for the preconcentration and/or separation of inorganic Cr species but hydrophobic organic Cr species have been isolated using a C_{18} column (Kaczynski and Kieber, 1994). Several methods were developed for use in the laboratory where the column effluent is directly transferred to the inlet of an appropriate analytical instrument, such as ICP or ICP-MS (Byrdy et al., 1995; Collins et al., 1997; Cox et al., 1985; Cox and McLeod, 1986). Other methods do not involve direct coupling and can be used to process samples either in the laboratory or in the field though, in the latter case, sample contamination may be an issue

(Beaubien et al., 1994; Cox and McLeod, 1992; Johnson, 1990). After reviewing these methods, the method of Beaubien et al. (1994) was chosen for preliminary evaluation.

The advantage of this method is that it was designed to separate Cr(III), Cr(VI), and colloidal Cr and to quantitate each of these fractions individually. In the method (described in detail in Chapter 2), a water sample is buffered with acetic acid and then passed through an anion-exchange resin that retains Cr(VI) and colloidal Cr but allows Cr(III) to pass through and be collected in the column eluant. Subsequently, Cr(VI) retained in the resin is reductively eluted by hydroxylamine HCl. Finally, colloidal Cr is eluted with a strong acid. Each eluent fraction is acidified and analyzed for total Cr.

A preliminary test of this method was conducted with two samples containing either 39.3 μg/L Cr(III) or 39.7 μg/L Cr(VI); triplicate sub-samples were processed. Performance of the method was acceptable with the sample containing only Cr (VI) though the average recovery was 120%. However, the method did not perform adequately when tested with the sample containing only Cr(III). Recovery of total Cr was 130% but only 77% was recovered in the Cr(III) fraction. A significant fraction of the total Cr was recovered in the Cr(VI) fraction. Results are reported in Table 4.4.

Table 4.4
Recovery of Cr species using the method of Beaubien et al. (1994)

Cr(VI)		Cr(III)		colloidal Cr		sum	
expected	measured	expected	measured	expected	measured	measured	% of expected
0	22.8	39.3	29.1	0	2.1	52.0	138
0	16.2	39.3	26.1	0	1.8	42.2	112
0	17.4	39.3	35.0	0	6.2	52.4	149
39.7	40.2	0	0.7	0	1.9	40.9	108
39.7	45.3	0	0.7	0	2.0	46.1	121
39.7	48.7	0	0.7	0	2.0	49.4	129

The poor performance of this method may have been partly due to a modification in which the size of the sample volumes processed was decreased from 500 mL in the published method to 10 mL. Although this volume reduction would have resulted in faster sample processing, it may have interfered with the effectiveness of the method. Further validation and optimization of this method would clearly be required prior to any field application.

During the course of this work, however, investigators from the U.S. Geological Survey tested a cation-exchange method for field speciation of Cr(VI) (Ball and McCleskey, 2003) and validated it against the EPA 218.6 ion chromatographic/colorimetric method (Arar and Pfaff, 1991). In the USGS method, Cr(III) is quantitatively retained on the cation exchange resin and the column eluent is analyzed for Cr(VI) as total Cr. Column eluent samples are preserved with acid but changes in the Cr oxidation state are of no concern since further analysis is done only for total Cr. Since this method addressed the concerns that were the focus of our study, further work on field methods for determining Cr redox speciation was not pursued.

DISSOLUTION OF Cr(III)-CONTAINING MINERALS

Dissolution studies were conducted using three types of geologic materials: the Cr-containing aluminosilicate mineral fuchsite, amorphous Cr(III) hydroxide, and aquifer sediments from the Cadiz site. Fuchsite was chosen as a model for an aquifer matrix mineral that could host (and release) Cr to groundwater. Significant deposits of fuchsite are found in Kern and San Bernardino Counties, California (Murdoch and Webb, 1966). Amorphous Cr(III) hydroxide, $Cr(OH)_3(s)$, was chosen as a model for Cr-containing oxide coatings that could form on aquifer minerals and to elucidate patterns of Cr release observed with fuchsite. These two model solids, which represent refractory (fuchsite) and labile ($Cr(OH)_3(s)$) sources of Cr, provide a basis for comparison with the aquifer sediment sample.

Since Cr occurs in groundwater as Cr(VI) but in most geologic materials as Cr(III), oxidation of Cr(III) to Cr(VI) must occur either co-incident with or subsequent to the release of Cr into solution. Here, hypochlorous acid/hypochlorite ($HOCl/OCl^-$) was used as a model oxidant to examine the oxidative dissolution of Cr(III)-containing solids. Kinetic experiments were conducted under both batch and flow-through conditions at acidic and basic pH. Under non-oxidizing conditions (examined for comparison), Cr(III) in solution was stabilized either by conducting experiments at low pH or, at high pH, by addition of the strong organic complexing agent ethylenediaminetetraacetate (EDTA).

Fuchsite Dissolution Reactions

Fuchsite is a type of muscovite, a dioctahedral 2:1 aluminosilicate mineral, that contains Cr(III). An ideal chemical formula of fuchsite can be expressed as:

$$K(Al_{2-x}Cr_x)(Si_3Al)O_{10}(OH)_2$$

where x indicates the stoichiometry of Cr in the solid.

Several reactions for fuchsite dissolution can be written depending on the reaction conditions. Under acidic conditions at pH ≤ 3.5,

$$K(Al_{2-x}Cr_x)(Si_3Al)O_{10}(OH)_2 + 10H^+ \rightarrow K^+ + xCr^{3+} + (3\text{-}x)Al^{3+} + 3H_4SiO_4 \quad (4.1)$$

When an oxidant (HOCl) is present in the solution, Cr(III) is oxidized to Cr(VI) and forms $HCrO_4^-$.

$$Cr^{3+} + \tfrac{3}{2}HOCl + \tfrac{5}{2}H_2O \rightarrow HCrO_4^- + \tfrac{3}{2}Cl^- + \tfrac{11}{2}H^+ \quad (4.2)$$

Therefore, the overall dissolution reaction in the presence of an oxidant becomes a combination of Eqns. 4.1 and 4.2.

$$\begin{aligned} &K(Al_{2-x}Cr_x)(Si_3Al)O_{10}(OH)_2 + \tfrac{3}{2}xHOCl + \tfrac{5}{2}xH_2O + (10 - \tfrac{11}{2}x)H^+ \rightarrow \\ &K^+ + (3-x)Al^{3+} + xHCrO_4^- + \tfrac{3}{2}xCl^- + 3H_4SiO_4 \end{aligned} \quad (4.3)$$

Eqns. 4.1 and 4.3 represent the dissolution reactions at low pH under non-oxidizing and oxidizing conditions, respectively, and are proton consuming reactions.

Under basic conditions at pH 9.0, dissolved Al and Cr form hydroxy complexes:

$$K(Al_{2-x}Cr_x)(Si_3Al)O_{10}(OH)_2 + (12\text{-}x)H_2O \rightarrow$$
$$K^+ + (3\text{-}x)Al(OH)_4^- + xCr(OH)^\circ_3 + 3H_4SiO_4 + (2\text{-}x)H^+ \quad (4.4)$$

In the presence of an oxidant, dissolved Cr(III) reacts to form $Cr^{VI}O_4^{2-}$ and the overall reaction becomes

$$\begin{aligned} &K(Al_{2-x}Cr_x)(Si_3Al)O_{10}(OH)_2 + \tfrac{3}{2}xOCl^- + (12 - \tfrac{3}{2}x)H_2O \rightarrow \\ &K^+ + \tfrac{3}{2}xCl^- + (3 - x)Al(OH)_4^- + xCrO_4^{2-} + 3H_4SiO_4 + (2 + x)H^+ \end{aligned} \quad (4.5)$$

In contrast to the reactions under acidic conditions, protons are released by the oxidative dissolution of fuchsite under basic conditions.

Elemental Composition of Fuchsite

Elemental compositions of fuchsite samples varied insignificantly among the five specimens, excluding F5 (Table 4.5). The result of the chemical analysis of sample F5 was not considered because the sum of weight % of the constituents was only 88.1 %, indicating that the data were not reliable. Based on the chemical analyses, the general chemical formula of fuchsite can be expressed as

$$K_{0.83}Na_{0.14}Mg_{0.01}Ca_{.0.01}Fe_{0.08}Mn_{0.001}Ti_{0.33}P_{0.01}Cr_{0.03}Al_{2.66}Si_{3.32}O_{12}$$

based on 12 oxygen atoms. Among the constituents, Si(IV) is the major tetrahedral cation with minor substitution by Al(III). Aluminum(III) is the major octahedral cation, with minor substitution by Fe(II)/Fe(III), Mn(II), Ti(IV), P, and Cr(III) in the octahedral sites. Potassium, Na^+, Mg^{2+}, and Ca^{2+} occur as interlayer cations. Silicon and Al are the dominant constituents, accounting for 44.7 ± 0.8 and 35.9 ± 2.3 mol %, respectively. Potassium is the next most abundant (11.1 ± 1.0 mol %) followed by Ti, Na and Fe in descending order. Chromium(III) constitutes 0.5 ± 0.2 mol % of the solid.

Dissolution Experiments with Fuchsite

The dissolution of fuchsite was studied in both batch and flow-through experiments. Batch experiments illustrate general trends in reactivity but steady-state dissolution rates are better characterized by flow-through experiments.

Batch dissolution experiments

In batch dissolution experiments, the presence of the oxidant resulted in a substantial increase in the amount of Cr released into solution at pH 9 but had a less dramatic effect at pH 2 (Fig. 4.1). This effect at pH 9 appears to be primarily associated with a rapid initial phase of the dissolution reaction. Steady-state rates are difficult to quantify in batch experiments because of the possibility of precipitation of secondary phases. However, the presence of EDTA, added to stabilize Cr(III) in solution at pH 9, had little observed effect on Cr release. Steady-state dissolution was further examined in flow-through experiments.

Table 4.5
Chemical composition (wt %) of fuchsite and aquifer sediment samples analyzed by XRF

	F1	F2	F3	F4	F5	F6	aquifer sediment Duplicate 1	aquifer sediment Duplicate 2
SiO_2	46.26	44.52	46.72	46.64	25.95	45.60	57.64	57.5
Al_2O_3	32.39	33.46	28.07	31.60	26.22	30.92	18.14	18.1
CaO	0.02	< 0.01	0.31	0.03	0.07	0.16	2.17	2.17
MgO	0.04	< 0.01	0.28	< 0.01	0.48	0.18	1.91	1.91
Na_2O	0.46	0.26	2.17	0.93	0.31	1.23	0.28	0.28
K_2O	9.90	9.36	7.75	8.95	6.10	8.82	7.47	7.49
Fe_2O_3*	1.40	0.93	1.73	1.06	15.15	2.51	6.74	6.74
MnO	0.01	< 0.01	0.01	< 0.01	0.76	0.05	0.04	0.04
TiO_2	5.03	6.09	8.23	6.15	1.62	5.15	0.73	0.73
P_2O_5	0.09	0.18	0.14	0.06	0.24	0.12	0.14	0.14
Cr_2O_3	0.71	0.48	0.33	0.49	9.44	0.96	0.01	0.01
LOI**	4.25	4.85	3.25	4.25	1.80	4.05	4.8	4.9
Sum	100.6	100.1	99.0	100.2	88.13	99.76	100.1	100.0

* as total Fe
** Loss on ignition

Flow-through dissolution experiments: Approach to steady state

In flow-through experiments, dissolution rates are calculated from the dissolved solute concentrations in the reactor effluent (Eqn. 2.1). The concentrations of Cr, Al, Si, and Fe in the reactor effluents vary during the course of the flow-through dissolution experiments as shown in Figures 4.2 and 4.3. Calculated dissolution rates were normalized to both the surface area and the elemental composition of the original solid. The normalization for elemental composition allows congruent and incongruent dissolution to be easily distinguished.

The normalized rates of Cr, Al, Si, and Fe release from fuchsite reached a constant value (i.e., steady state) by about 12 d (279 ± 50 h) under all conditions (Fig. 4.4). Steady-state dissolution is expected under conditions where: (1) solution composition is constant over time, (2) the solid is not substantially depleted during the reaction, and (3) secondary phases are not precipitated. A period of rapid, initial dissolution is commonly observed to precede the establishment of a steady-state dissolution rate (Stumm, 1992). Such transient rapid dissolution is usually attributed to the dissolution of fine (submicron) particles adhering to the solid (Berner and Holdren, 1979; Eggleston et al., 1989) or other experimental artifacts. Although interpretation of initial dissolution rates is questionable, it is notable that the extent and magnitude of the rapid initial dissolution varied with reaction conditions and among the different constituents of the solid.

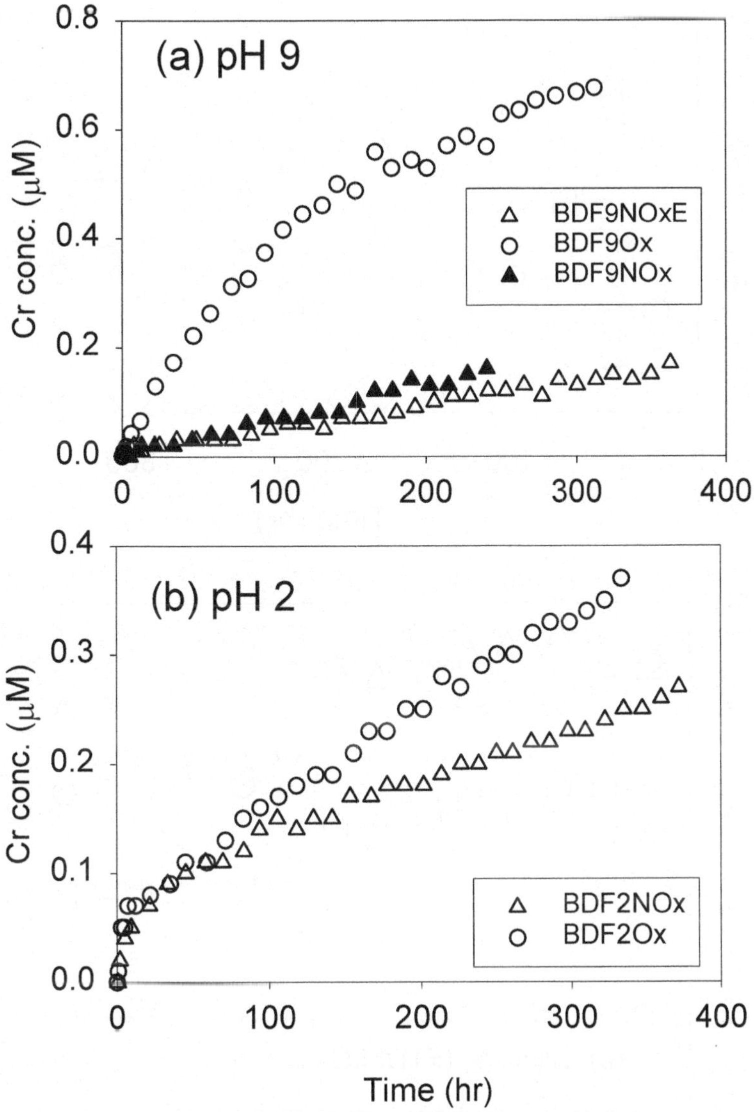

Figure 4.1 Concentrations of dissolved Cr over time measured in samples from batch dissolution of fuchsite at pH 9 and 2. Experimental conditions are given in Table 2.1.

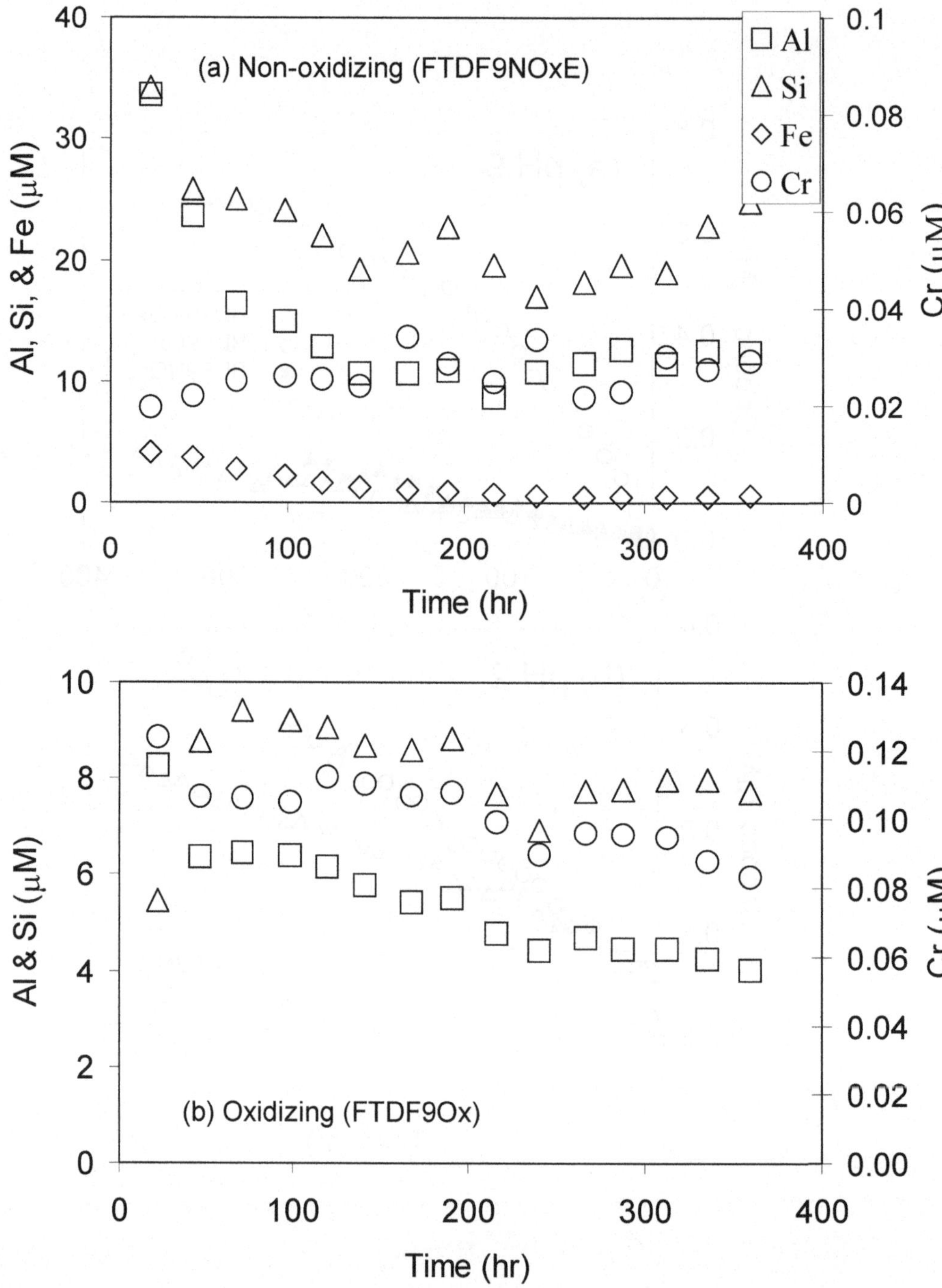

Figure 4.2 Concentrations of dissolved metals released from fuchsite determined by flow-through experiments at pH 9. Experimental conditions are given in Table 2.3.

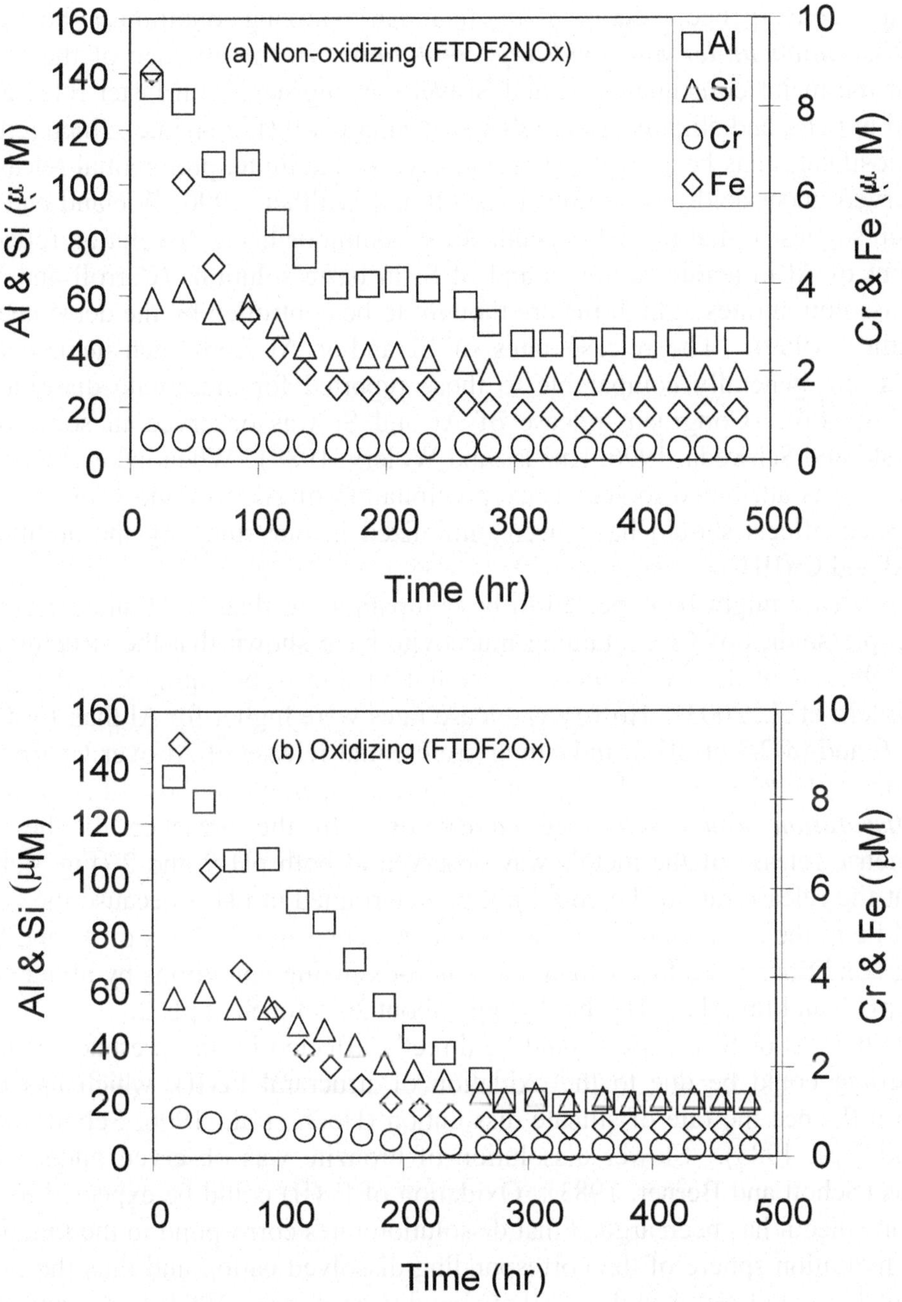

Figure 4.3 Concentrations of dissolved metals released from fuchsite determined by flow-through experiments at pH 2. Experimental conditions are given in Table 2.3.

Flow-through dissolution experiments: Steady-state dissolution

As shown in Figure 4.5, the normalized steady-state rates of metal release from fuchsite show greater convergence under oxidizing than non-oxidizing conditions at both pH 2 and 9.

Dissolution under non-oxidizing conditions. In the absence of the oxidant, even the release of the major constituents Al and Si were incongruent with faster release (by a factor of 1.7) of Al at pH 2 and slightly faster release of Si at pH 9 (Fig. 4.5a,c). Incongruent dissolution of aluminosilicates has been reported previously; for kaolinite, preferential release of Al over Si was observed under acidic conditions (Carroll and Walther, 1990; Wieland and Stumm, 1992). It has been suggested that the pH dependence of aluminosilicate dissolution reflects rate limiting detachment of Al in acidic solutions and of Si in basic solutions (Carroll and Walther, 1990). Silicate dissolution rates at high pH are thought to be controlled by the detachment of Si (Brady and Walther, 1989). The release rates of Si and/or Al from fuchsite under non-oxidizing conditions, are generally comparable to those reported for muscovite dissolution in previous studies (Fig. 4.6). Congruent release of Al and Si was observed in some of these studies (Kalinowski and Schweda, 1996; Knauss and Wolery, 1989). When incongruent dissolution was observed, it was attributed to secondary precipitation of Al (Lin and Clemency, 1981; Nickel, 1973). Such effects should have been minimized in our study by the addition of EDTA to chelate Al and Cr(III).

Cr release might be expected to be comparable to that of Al since crystallographic and spectroscopic studies of Cr-containing muscovite have shown that the structural Cr is trivalent and that >99.5 % of the Cr occupies octahedral sites by substituting for Al(III) (Brigatti et al., 2001; Cardelli et al., 2003). However, release rates were higher for Al than for Cr by a factor of 10 at pH 9 and of 2.3 at pH 2, indicating preferential release of Al over Cr from the solid. The release rates of Fe, also an octahedral cation, were comparable to those of Al at both pH 9 and 2.

Dissolution under oxidizing conditions. In the presence of the oxidant, nearly stoichiometric release of the metals was observed at both pH 9 and 2 (Fig. 4.4). It should be noted that the release rate of Fe could not be determined at pH 9 because the concentrations of dissolved Fe in the samples were all below the detection limit. Rates of Al and Si release under oxidizing conditions were lower than under non-oxidizing conditions by about 50% for both Al and Si at pH 9 and for Al at pH 2 but by only about 20% for Si at pH 2.

Neither Al or Si release should be directly affected by the presence of the oxidant. An indirect effect could be due to the oxidation of structural Fe(II), which has previously been shown to influence the rate of silicate dissolution (Hoch et al., 1996; Schott and Berner, 1983; White and Yee, 1985). Slower dissolution of bronzite was observed under oxic than anoxic conditions (Schott and Berner, 1983). Oxidation of Fe(II) could be expected to result in slower dissolution since it has been argued that dissolution rates correspond to the kinetics of water loss from the hydration sphere of the corresponding dissolved cation and thus the Fe(III)-O bond is inherently slower to react than the Fe(II)-O bond (Casey et al., 1993; Casey and Westrich, 1992). In contrast, release of Si from pyroxene was found to be enhanced at higher dissolved O_2 concentrations (Hoch et al., 1996). This would be consistent with a proposed mechanism for oxidative dissolution as a coupled electron-cation transfer in which charge balance in the solid would require enhanced release of a cation upon oxidation of structural Fe(II) (White and Yee, 1985). At this point, the effect of the oxidation of a trace constituent of a solid on the release rates of the major constituents is not clearly understood.

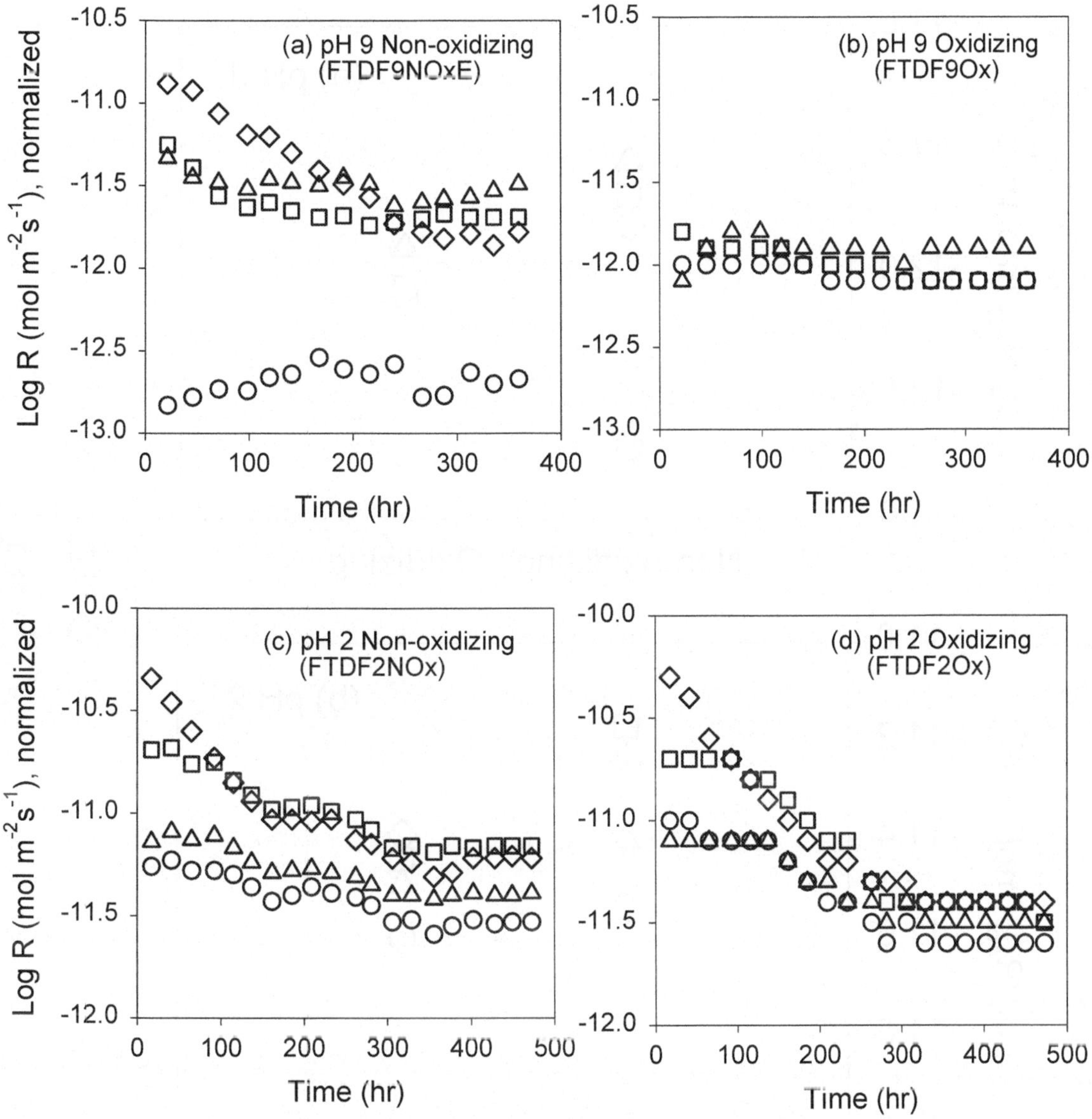

Figure 4.4 Logarithm of the release rates of metals from fuchsite determined by flow-through experiments. Release rate of each metal was normalized by the content of the metal in the original solid determined by XRF analyses. Experimental conditions are given in Table 2.3.

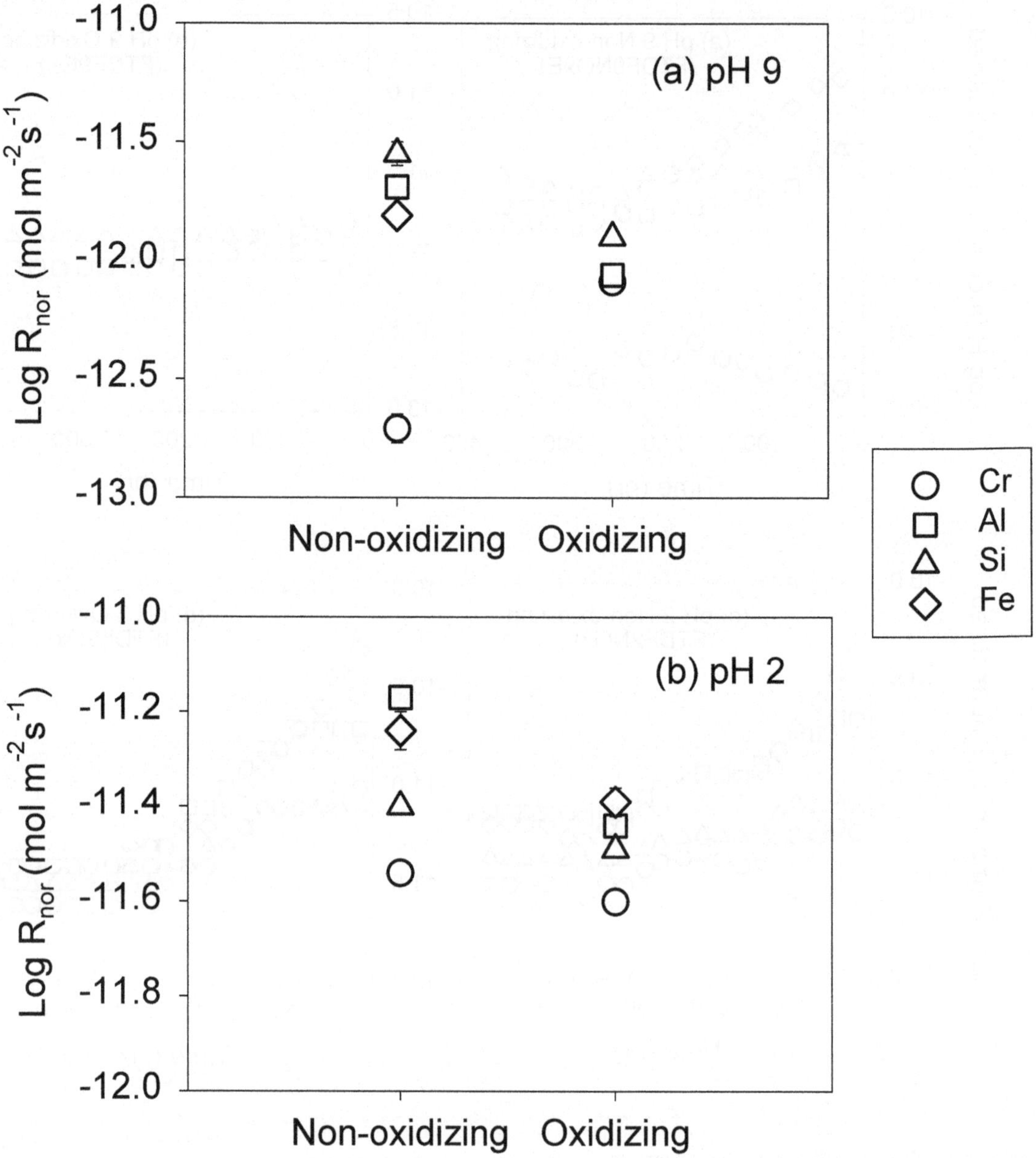

Figure 4.5 Normalized release rates of metals from fuchsite after a steady-state was reached under oxidizing or non-oxidizing conditions. Experimental conditions are given in Table 2.3.

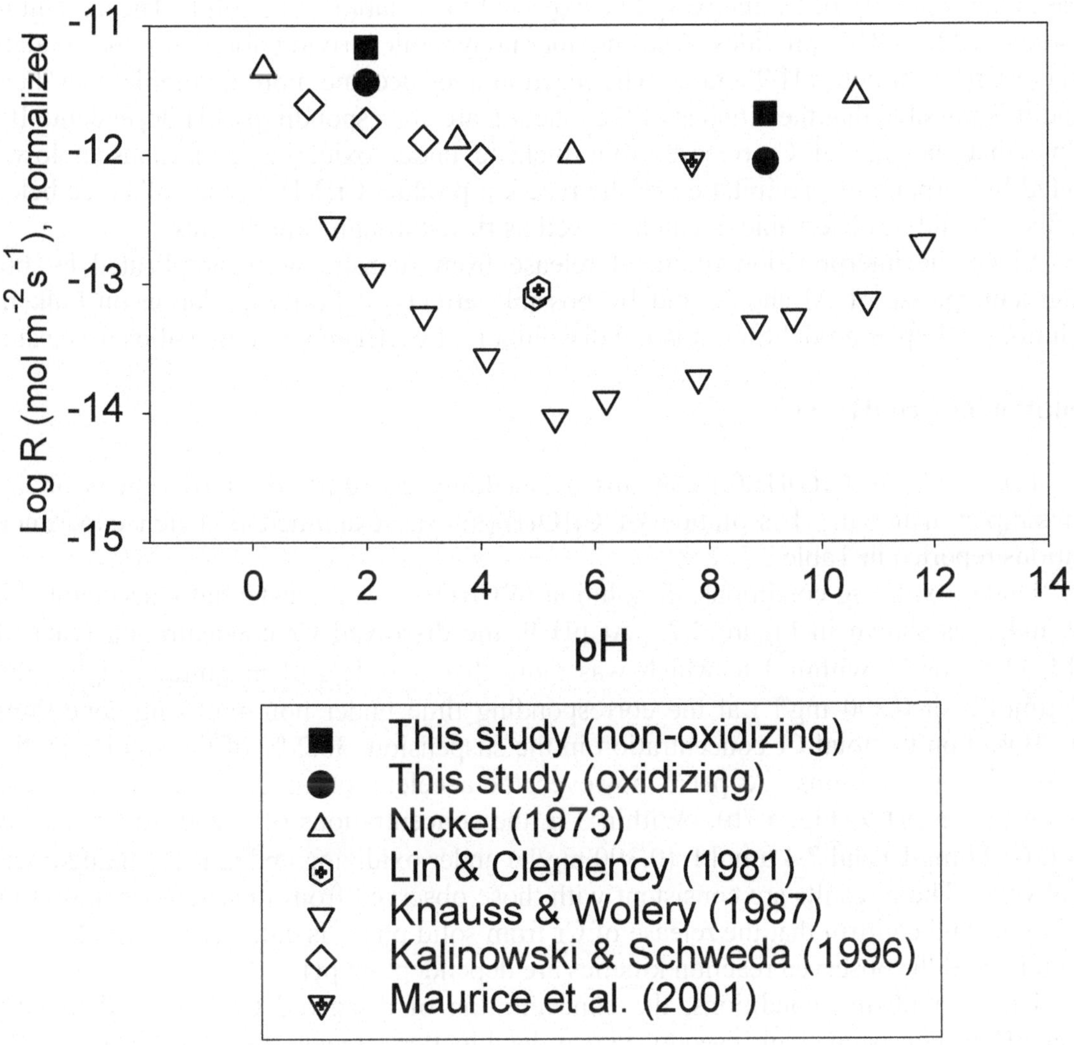

Figure 4.6. Dissolution rates of fuchsite and muscovite under various conditions. Rates are based on Al release for this study and Si release for all others. Some values for log R (Knauss and Wolery, 1989; Maurice et al., 2002) **were extrapolated to 25 °C using an activation energy (E_a) of 14 kcal/mol** (Lasaga, 1981).

The rate of Cr release was increased under oxidizing condition by a factor of 3.7 at pH 9 but decreased by about 13% at pH 2. The acceleration at pH 9 was expected based on the results of the batch studies (Fig. 4.1) but not the deceleration at pH 2. Oxidation of Cr(III) to Cr(VI) increases the solubility of Cr and would be expected to destabilize the solid. The oxidant used in this study, $HOCl/OCl^-$, provides a strong thermodynamic driving force for the oxidation of Cr(III) to Cr(VI) at both pH 9 and 2. The reaction does become more favorable with increasing pH and it is possible that the kinetics of the reaction are correspondingly pH dependent. It is also possible that the rate of Cr release from fuchsite under oxidizing conditions at low pH is inhibited by sorption or precipitation of the reaction product Cr(VI) on the solid (see below) but this effect should be observable in batch as well as flow-through experiments.

Since the interpretation of metal release from fuchsite was complicated by both the incongruent release of Al and Si and by possible effects of Fe(II) oxidation on bulk mineral dissolution, a simpler model for oxidative dissolution of Cr(III)-containing solids was examined.

Dissolution of $Cr(OH)_3(s)$

Like fuchsite, $Cr(OH)_3(s)$ contains octahedrally coordinated Cr(III) but is obviously a much simpler material. Dissolution of $Cr(OH)_3(s)$ was examined in batch experiments for conditions reported in Table 2.2.

Under oxidizing conditions, dissolution of $Cr(OH)_3$ was substantially accelerated at both pH 9 and 2 as shown in Figure 4.7. At pH 9, the dissolved Cr concentration reached 3720 μmol/L (195 mg/L) within 3 h, which was more than 2 orders of magnitude higher than that (15.2 μmol/L or 0.800 mg/L) at the corresponding time under non-oxidizing conditions (Fig. 4.7a). Based on the total Cr concentration in the suspension, 38.2 % of the solid was dissolved under oxidizing conditions. At pH 2, however, the acceleration of Cr release by oxidation was less than that at pH 9 (Fig. 4.7b). Within 3 h, the concentrations of dissolved Cr reached 80.7 μmol/L (4.24 mg/L) and 7.4 μmol/L (0.390 mg/L) under oxidizing and non-oxidizing conditions, respectively. These results are consistent with those observed from fuchsite dissolution in batch experiments and confirm that the release of Cr from solid phase is enhanced by oxidation at both pH 9 and 2 and the observed reaction kinetics are dependent on pH.

Under oxidizing conditions, the concentration of dissolved Cr increased linearly with time at pH 9, consistent with pseudo zero order kinetics. Generally, the dissolution kinetics follow a zero-order rate expression under steady state conditions (Stumm and Wollast, 1990). At pH 2, however, the release rate of Cr rapidly decreased with time and the concentration of dissolved Cr reached a plateau (Fig. 4.7b). This result indicates that the dissolution of the solid is inhibited and the inhibitory effect increases with time. Under oxidizing conditions, the product of the dissolution reaction, Cr(VI), can adsorb on the residual solid, particularly in acidic solutions. If the adsorption of Cr(VI) on the solid is substantial, it might inhibit the dissolution by blocking the reactive surface sites of the solid, as is discussed in the following section.

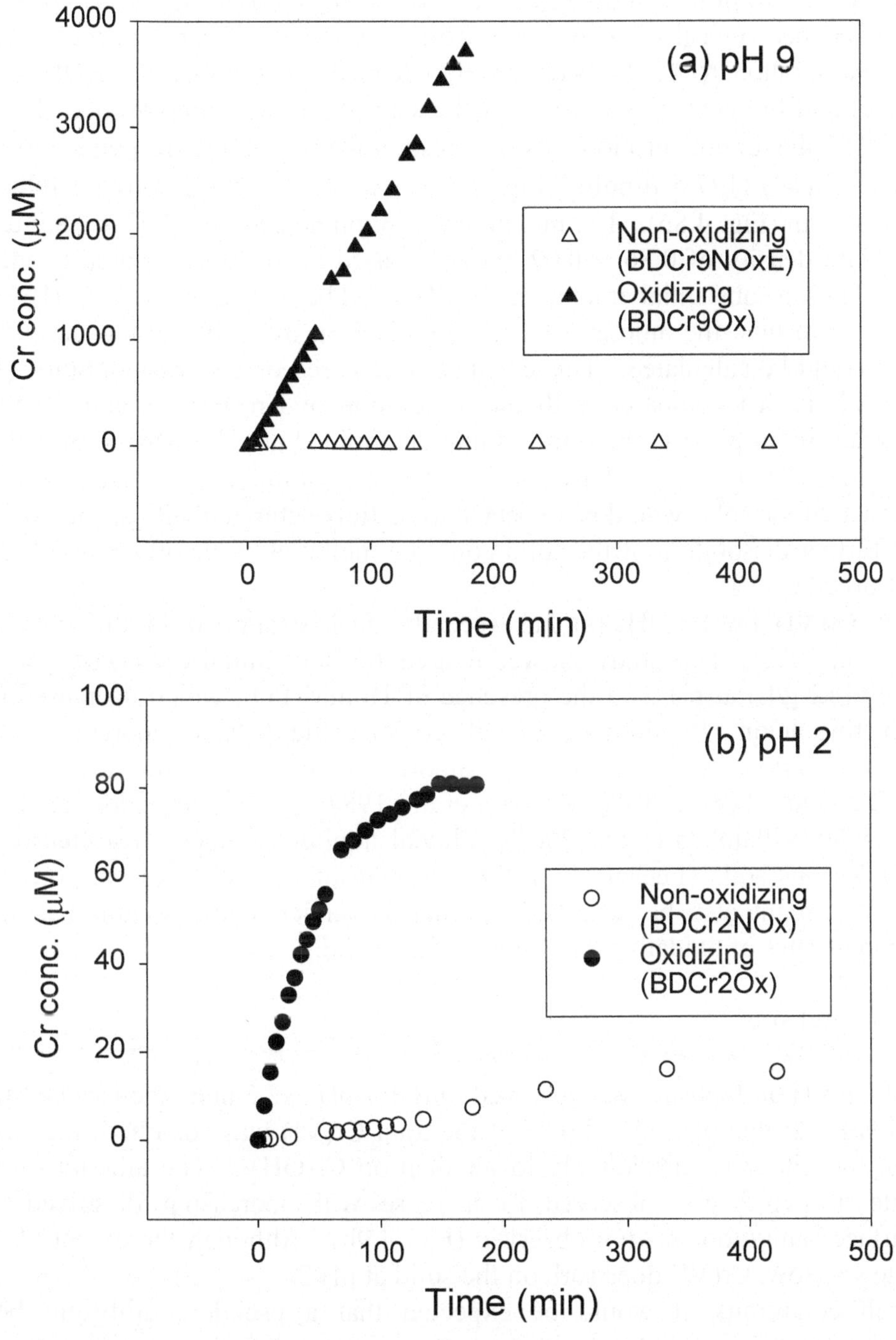

Figure 4.7 Dissolved Cr concentrations released by dissolution of $Cr(OH)_3(s)$ at pH 9 and 2. Experimental conditions are given in Table 2.2.

Cr(VI) Sorption on $Cr(OH)_3(s)$

The extent of Cr(VI) sorption on $Cr(OH)_3(s)$ was measured as a function of both the pH and variable Cr(VI) concentration as shown in Figure 4.8. The sorption of Cr(VI) on the solid exhibited a pH dependence typical of anions, for which the extent of sorption increases with decreasing pH. Sorption onto $Cr(OH)_3(s)$ is expected to be analogous to that on $Fe(OH)_3(s)$ (i.e., hydrous ferric oxide) and to occur by the same mechanism of surface complexation (Dzombak and Morel, 1990). The maximum sorption was observed at pH ~ 3 and corresponded to uptake of 92 % of the total Cr(VI) (177.5 μmol/L (Fig. 4.8a). At pH 3, Cr(VI) sorption followed a typical Langmuir isotherm (Fig. 4.8b). The maximum sorption density was 307.8 ± 8.3 μmol/g, corresponding to a site density of 0.86 ± 0.02 site/nm^2 based on the surface area of the solid (215.0 ± 1.1 m^2/g). Thus, a substantial amount of Cr(VI) could be sorbed on the $Cr(OH)_3$ solid.

Based on these results, the amount of Cr(VI) sorbed on the solid during the oxidative dissolution at pH 2 could be calculated. This calculation is based on two assumptions: (1) that the sorption of Cr(VI) is fast compared with the dissolution of $Cr(OH)_3(s)$, and (2) that the maximum sorption density at pH 2 is the same as that determined at pH 3. Since the dissolved Cr concentration reached 80.7 μmol/L during the oxidative dissolution of $Cr(OH)_3(s)$ at pH 2 (Fig. 4.7b), the surface of the solid would be expected to be fully saturated with adsorbed Cr(VI). This suggests that further dissolution of the solid could be inhibited by the blocking of reactive surface sites by adsorbed Cr(VI).

Sorption of Cr(VI) on $Cr(OH)_3(s)$ has recently been examined in the presence of phosphate (Tzou et al., 2003). This study showed that ca. 60 % of initial Cr(VI) (0.4 mM) was adsorbed on the solid (2 g/L) at pH 3 in the presence of 10 mM PO_4, which decreased Cr(VI) sorption as a competing anion. In addition, several previous studies have reported that Cr(VI) sorption occurred on various sorbents such as amorphous iron oxyhydroxide (Aoki and Munemori, 1982; Davis and Leckie, 1980; Zachara et al., 1987), goethite (Fendorf et al., 1997; Van Geen et al., 1994; Villalobos et al., 2001), alluvial aquifer sediments (Stollenwerk and Grove, 1985), and various soils (Lee et al., 1999). The results of this and previous studies suggested that Cr(VI) might also be adsorbed on fuchsite under acidic conditions, possibly depressing the apparent rates of oxidative dissolution of the solid.

Cr(VI) Sorption on Fuchsite

Sorption of Cr(VI) on fuchsite was observed only for pH ≤ 2.5 and reached 0.02 μmol/g at pH 1.95 (Fig. 4.9a). At this pH, only 4.6 % of the total Cr(VI) was adsorbed; the extent of Cr(VI) sorption on fuchsite was substantially lower than on $Cr(OH)_3$. The amount of Cr(VI) sorbed on fuchsite at pH 2 was observed to increase with increasing dissolved Cr(VI) concentrations; surface saturation was not observed (Fig. 4.9b). Although the extent of Cr(VI) sorption on fuchsite was low, Cr(VI) does sorb on the solid at pH 2.

Under flow-through conditions, it would be expected that a pseudo-equilibrium between dissolved and adsorbed Cr(VI) would be established even as the dissolution reaction progressed. The flux of dissolved Cr from the flow-through reactor should thus correspond to the rate of fuchsite dissolution (as in the case without adsorption of the reaction product). At pH 2 under oxidizing conditions, the steady-state dissolved Cr concentration was 0.20 μM (Fig. 4.3b). The corresponding pseudo-equilibrium concentration of adsorbed Cr(VI) would be 0.004 μmol/g (cf.

Fig. 4.9b) or 0.07 μM at the solid concentration in the flow-through experiment. This sorbed concentration is clearly well below surface saturation.

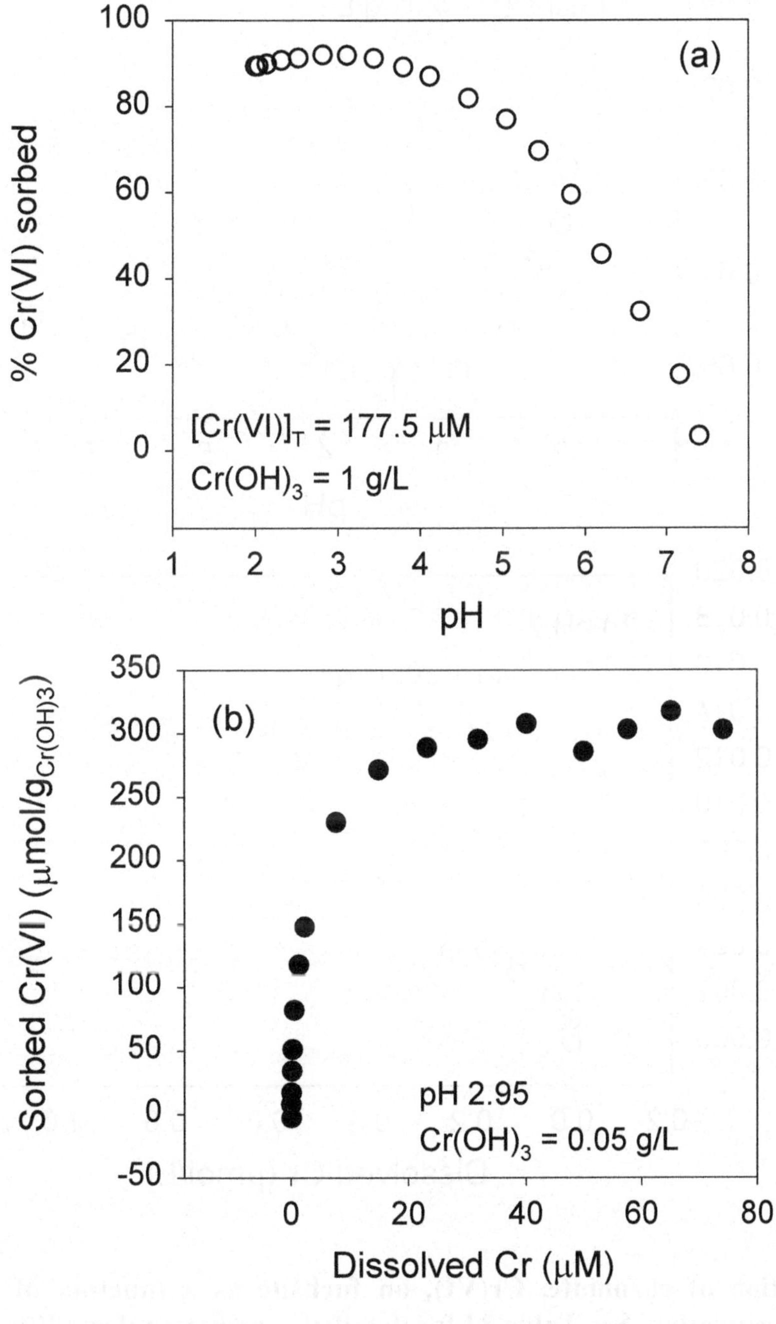

Fig. 4.8 Sorption of chromate, Cr(VI), on $Cr(OH)_{3(s)}$ as a function of (a) pH and (b) dissolved Cr concentration. Experimental conditions are given in Table 2.5.

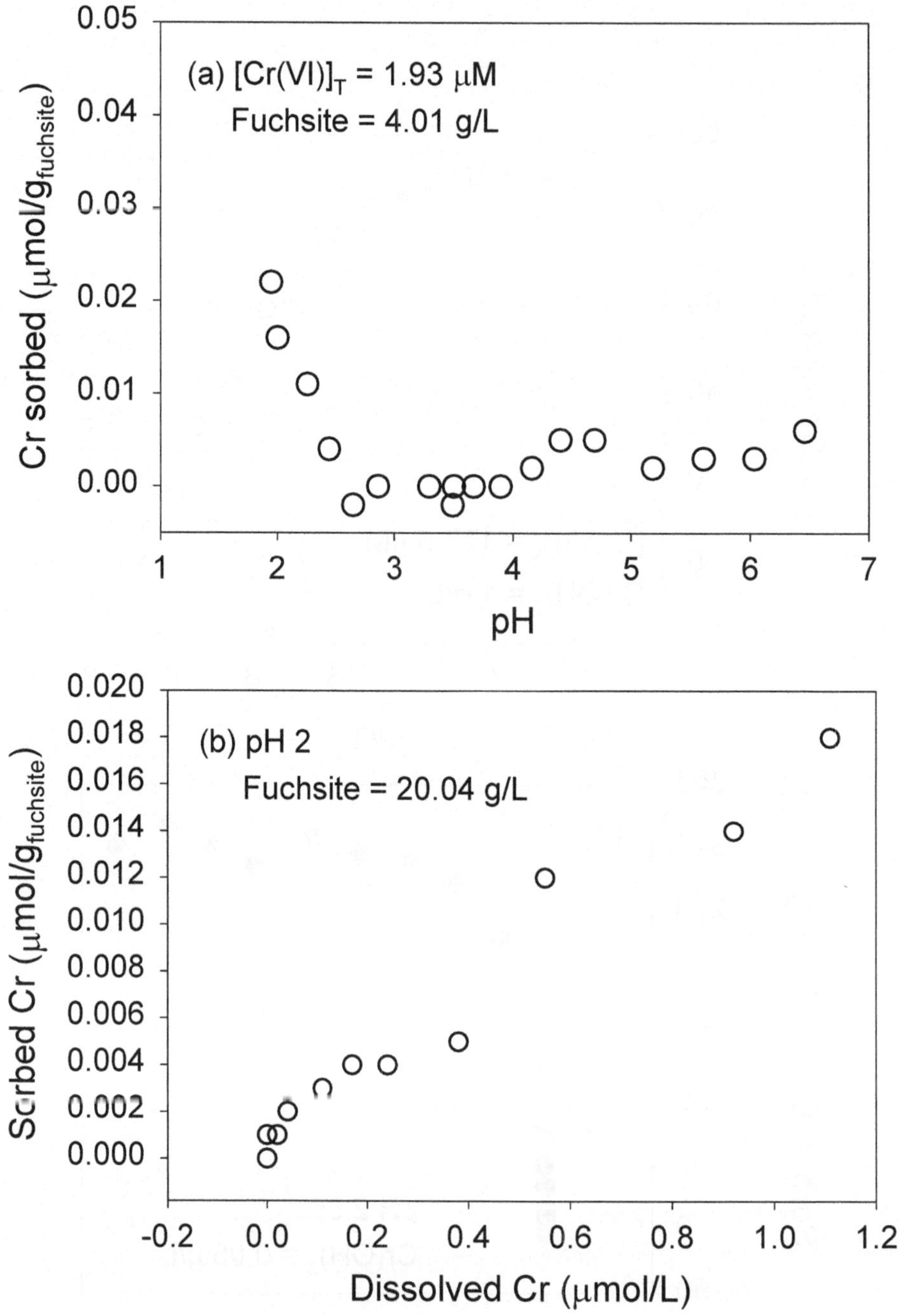

Figure 4.9 Sorption of chromate, Cr(VI), on fuchsite as a function of (a) pH and (b) dissolved Cr concentration. See Table 24 for detailed experimental conditions.

Dissolution of Aquifer Sediment

The aquifer sediment sample (Cadiz drill-cutting OW-2) consisted mostly of Si and Al, which accounted for 57.6 % and 18.1 % by weight, respectively (Table 4.5). The Cr content was 0.01 wt%, which is the analytical detection limit of this element. Under non-oxidizing conditions at pH 9, hardly any dissolution of the solid was observed. The concentrations of dissolved Si and Al were not observed to increase with time over 291 h but rather remained constant at 2.2 ± 0.2 and 2.8 ± 1.1 μM, respectively. Dissolved Cr concentrations were always below the detection limit (0.01 μM).

Under oxidizing conditions, however, release of metals from the solid was observed. In a preliminary experiment, initiated with dry sediment, no dissolution was observed for 6 d but, after that period, the concentrations of dissolved Cr, Al, and Si began to increase rapidly. Such behavior is unusual in dissolution experiments and may be attributable to the solid having been stored dry and at uncontrolled (but likely elevated) temperature for an extended period. Therefore, in a subsequent experiment, the sediment was pre-equilibrated with the (pH 9, non-oxidizing) buffer solution for 2 d and rinsed with the buffer 5 times before the oxidative dissolution was initiated by addition of the oxidant. In this experiment, no lag period was observed before the release of metals into solution (Fig. 4.10). In the presence of the oxidant, Cr was released from the aquifer sediments even though the Cr content of the solid was very low. The concentrations of dissolved Cr increased linearly with time and reached 0.08 μM (4.3 μg/L or 0.43 μg/g) in 14 d. The dissolved concentrations of Al and Si increased rapidly over the first 2 d, after which they were released at a slower rate. Dissolved Al and Si concentrations reached 8.6 μM (231.7 μg/L) and 43.5 μM (1221.9 μg/L), respectively, during the dissolution experiment. These results indicate that the oxidation of Cr(III) could be an important factor controlling the release of Cr from aquifer sediments into groundwater.

WATER QUALITY ISSUES IN AQUIFER STORAGE AND RECOVERY

The release of metals from aquifer sediments plays a key role in the evolution of groundwater quality under conditions of both natural and artificial recharge. This process, as well as sorption-desorption and the precipitation of secondary minerals, are examined both in the general case and in reference to the oxidative dissolution of Cr(III)-containing minerals.

Conceptual Model for Evolution of Recharge Water Composition

The key aspect of the evolution of the chemical composition of recharge water is the partitioning of chemical constituents between the mobile phase (dissolved and mobile colloids) and the immobile phase (solid and immobile colloids). Mobile constituents pose the most direct threat of human exposure since they can migrate from their source region to a drinking water well. The type of process by which a constituent, X, may be mobilized (or sequestered) depends on its speciation in the solid phase. If X is an integral constituent of an aquifer mineral, dissolution of the mineral will be required to release X into solution. Whereas, mineral dissolution would not be required to mobilize X if it is sorbed onto a mineral surface. If X is present in association with colloidal particles, its mobility will be determined by processes of colloidal attachment to or detachment from the immobile phase. Environmental conditions (including both chemical composition of pore fluids and microbial activity) will influence the extent of mobilization or sequestration of X. For metals, in particular, both solubility and the affinity for surfaces can be strongly influenced by redox conditions and the presence of (biogenic) complexing agents.

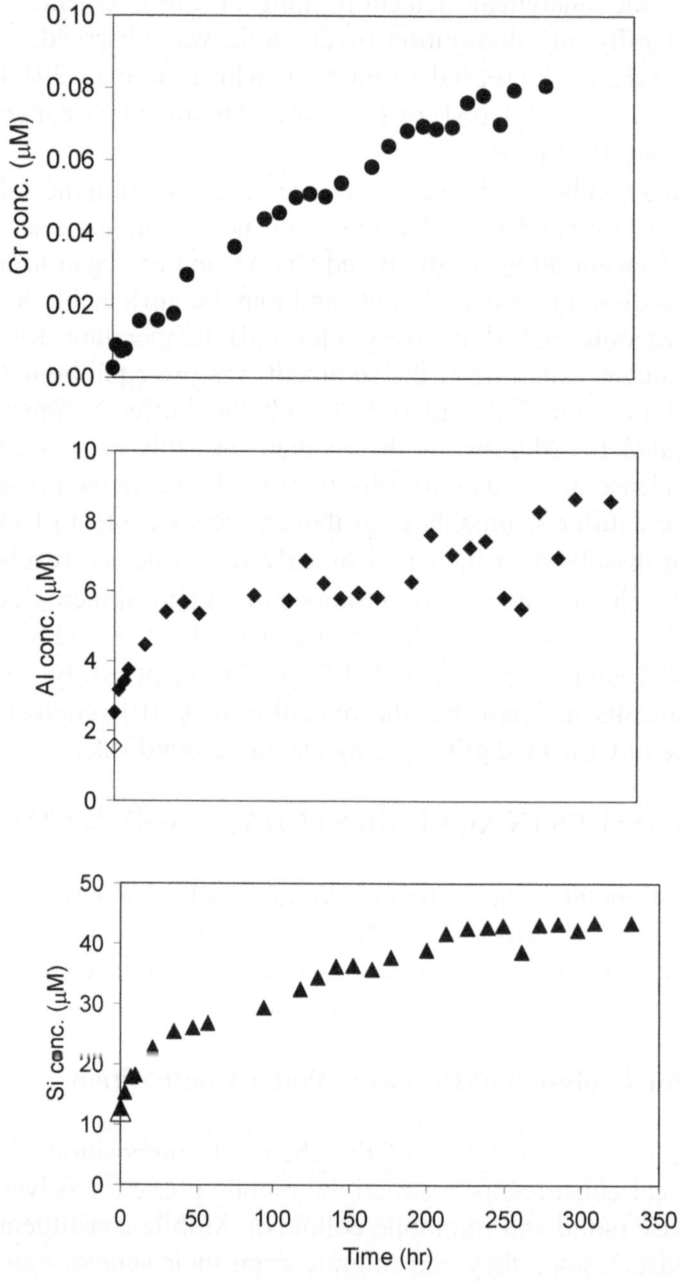

Figure 4.10 Release of metals from aquifer sediments during the oxidative dissolution at pH 9. Conditions: 10.0 g/L solid (Cadiz aquifer drill-cuttings, OW-2), 0.01 M $NaNO_3$, 0.01 M $Na_2B_4O_7$, 1 mM NaOCl, and pH 9.07 ± 0.09. Open symbols indicate the concentrations of dissolved Al and Si in the fifth washing solution. Solid was pre-equilibrated with buffer solution for 2 d.

A schematic diagram illustrating the partitioning of X between mobile and immobile phases by dissolution of a primary solid, precipitation-dissolution of a secondary solid and sorption-desorption onto aquifer minerals is shown in Figure 4.11. These processes, which would lead to the accumulation of X in the recharge water along a flowpath, are influenced by a variety of factors, including the primary and secondary minerals present in the soil or aquifer matrix and the composition of the infiltrating water. The presence and concentrations of co-constituents with X in the recharge water will affect its corrosiveness toward soil and aquifer minerals and the extent of saturation of X with respect to secondary minerals or competition with X for sorption sites on mineral surfaces.

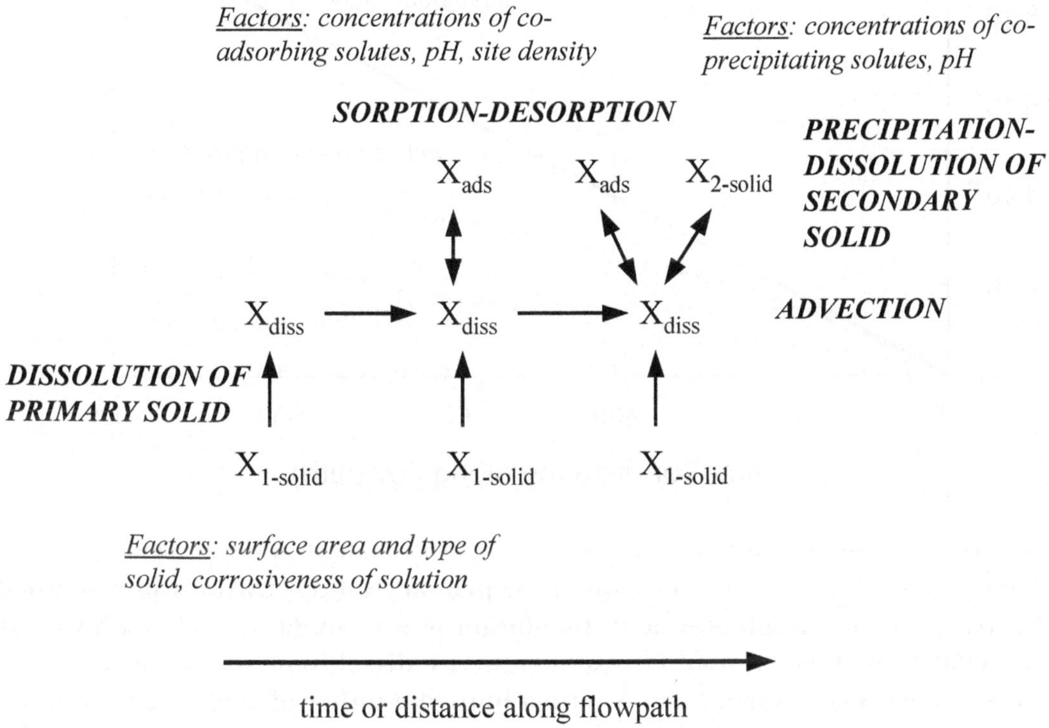

Figure 4.11 Schematic of processes leading to partitioning of constituent X between the mobile phase (i.e., dissolved X, X_{diss}) and immobile phases (i.e., X as a constituent of primary or secondary minerals, $X_{1\text{-solid}}$ or $X_{2\text{-solid}}$, or X sorbed onto mineral surfaces, X_{ads}).

A simple model based on these processes describes the concentration of constituent X in solution with time in a batch reactor (corresponding to distance along a flowpath) as shown in Figure 4.12. In this model, the primary solid is considered as an infinite source of X that undergoes irreversible dissolution. The secondary solid is allowed to precipitate (reversibly) thus altering the dissolved concentration of X. X is also allowed to sorb onto (and desorb from) surface sites on the aquifer minerals, which are assumed to be limited (i.e., saturable). (Note that a linear (i.e., K_D model) for sorption, which is often applied in modeling contaminant transport in groundwater, has an effect comparable to (reversible) precipitation-dissolution of a secondary phase.)

As shown in Figure 4.12, the increase in the concentration of dissolved X with time is modified by the inclusion of reversible precipitation and sorption processes. When reversible precipitation-dissolution of a secondary solid is allowed, the dissolved X concentration eventually reaches a steady-state value that is determined by pseudo-equilibrium with the secondary solid (which is reached even as dissolution of the primary solid continues). With the inclusion of sorption-desorption processes, the accumulation of dissolved X is slowed but because sorption sites are limited, the (eventual) steady-state concentration of dissolved X is unchanged.

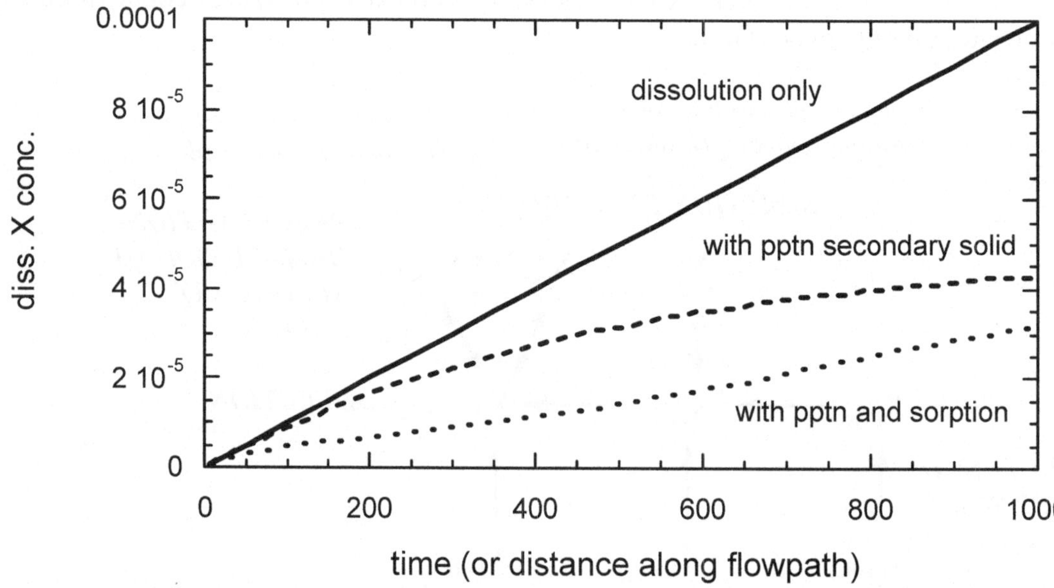

Figure 4.12 Predicted dissolved concentrations of X with time in a batch reactor (or with distance along a flowpath). (—) Dissolution of the primary solid releases X into solution at a constant rate. (– – –) Precipitation and dissolution of a secondary solid is allowed such that the secondary solid is in pseudo-equilibrium as dissolution of the primary solid continues. (- - -) Sorption-desorption of X onto a limited number of surface sites is allowed together with precipitation-dissolution of the secondary solid and dissolution of the primary solid. Note: units are arbitrary.

Sorption Effects Illustrated in a Model System

The oxidative dissolution of $Cr(OH)_3(s)$ in batch experiments (presented above) provides a useful model to examine the effects of sorption on the rates and extent of dissolution. Experiments were conducted at pH 2 and 9 with and excess of the strong oxidant hypochlorous acid/hypochlorite ($HOCl/OCl^-$). Under both pH conditions, the reactions

$$\text{(pH 2)} \qquad 2Cr(OH)_3(s) + 3HOCl \rightarrow 2HCrO_4^- + 3Cl^- + H_2O + 5H^+$$

$$\text{(pH 9)} \qquad 2Cr(OH)_3(s) + 3OCl^- \rightarrow 2CrO_4^{2-} + 3Cl^- + H_2O + 4H^+$$

are strongly favored thermodynamically.

At pH 9, linear dissolution kinetics are observed over the reaction time course of about 200 min during which approximately 40% of the solid was oxidatively dissolved (Fig. 4.7a). Over the same time period at pH 2, substantially less Cr was released into solution and the rate of the reaction decreased markedly (Fig. 4.7b). In both cases, the presence of the oxidant significantly increased the extent of solid dissolution. The pH dependence of the oxidative dissolution reaction corresponds to the sorption of the product Cr(VI) onto $Cr(OH)_3(s)$, which is negligible at pH 9 and reaches a maximum at pH 2-3 (Fig. 4.8a). At pH 3, Cr(VI) sorption is limited by the surface site concentration and reaches a maximum sorption density of 16,000 μg/g (Fig. 4.8b).

Clearly sorption of Cr(VI) can influence the extent to which the reaction product appears in solution. Since determination of the rate of oxidative dissolution is based on the release of Cr to solution, sorption could lead to an underestimate of the dissolution rate. This artifact, however, will be insignificant once the surface sites are saturated. Thus, in order to explain the observed pH dependence of the oxidative dissolution rate, we hypothesize that sorption of Cr(VI) at the $Cr(OH)_3(s)$ surface inhibits the further reaction of the solid. (Note that it is possible that the rate of oxidative dissolution is inherently pH dependent but, for illustrative purposes, we do not include this effect.)

In a simple model that incorporates a pseudo-zero order oxidative dissolution of $Cr(OH)_3(s)$ modified by the fraction of free surface sites, the change in the dissolved Cr(VI) concentration, $[Cr(VI)]_{diss}$, over time is given by Eqn. 4.6:

$$\frac{d[Cr(VI)]_{diss}}{dt} = k_{ox}\left(\frac{[S]}{[S]_T}\right)^n - k_{ads}[S][Cr(VI)]_{diss} + k_{des}[Cr(VI)]_{ads} \qquad (4.6)$$

where k_{ox}, k_{ads}, and k_{des} are the rate constants for oxidative dissolution, adsorption, and desorption respectively, [S] and $[S]_T$ are the free and total surface site concentrations, $[Cr(VI)]_{ads}$ is the adsorbed Cr(VI) concentration, and n is an exponent of arbitrary value.

As can be seen in Figure 4.13a, sorption has little effect on the observed concentration of dissolved Cr(VI) because this concentration rapidly exceeds that needed to saturate the surface. However, if the sorbed Cr(VI) is allowed to inhibit further reaction, the amount of Cr released into solution is dramatically decreased and the reaction rate decreases sharply (Figure 4.13b). Even though the initial reaction rate (i.e., without inhibition) is identical in both cases, the adsorption of dissolved Cr(VI) initially prevents its accumulation in solution and inhibition by adsorbed Cr(VI) then effectively stops the reaction. It should be noted, however, that the predicted values are sensitive to the values chosen for the rate constants for adsorption and desorption and the strength of the inhibitory effect (i.e., the value of n). The value for k_{ox} in this model was chosen to match the observed rate of Cr release at pH 9. Although the model was not further optimized, the predictions suggest that the same (i.e., a pH-independent) value of k_{ox} could be consistent with the observations at pH 2 if an inhibitory effect of sorbed Cr(VI) is included. However, it is possible that precipitation of a Cr(VI)-containing solid, rather than Cr(VI) sorption, could be responsible for the observed inhibition of oxidative dissolution.

Comments

In interpreting the occurrence of naturally-occurring Cr(VI) in groundwater, the geologic source of Cr and the processes by which Cr is released into the infiltrating water and oxidized from Cr(III) to Cr(VI) must be considered. Although Cr(VI) is thermodynamically favored in oxic waters over a wide pH range and the crustal abundance of Cr is high for a trace element, the occurrence of Cr(VI) in groundwater is limited by one or more of the following processes: slow reaction of Cr(III) in solid phases with dissolved oxygen, slow (non-oxidative) release of Cr(III) from solid phases, precipitation of Cr(III) as $Cr(OH)_3(s)$, and/or availability and reactivity of Mn(III, IV) under aquifer conditions. The elevated pH typical of groundwaters in arid regions favor the mobility of Cr(VI) by inhibiting sorption.

Cr(VI) provides a specific example of the more general phenomena of water-rock interactions that govern the evolution of the chemical composition of recharge water. The processes of dissolution, precipitation, and sorption can both introduce and remove chemical constituents from the infiltrating water. Chemical constituents released by dissolution can undergo other reactions (e.g., precipitation or sorption) and these reactions may inhibit further dissolution. The composition of recovered water in a groundwater storage project will then reflect the type and properties of soil and aquifer minerals, the extent and duration of the water-rock contact, and the corrosiveness of the recharge water toward the aquifer minerals.

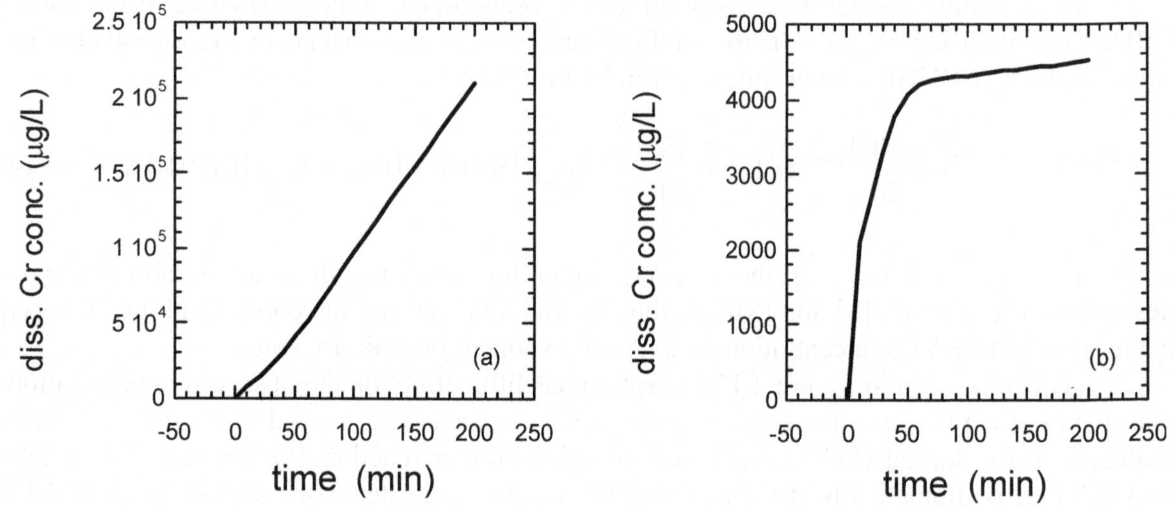

Figure 4.13 Predicted concentrations of Cr(VI) in solution over time based on the model as described in the text with k_{ox} = 1130 μg $L^{-1}min^{-1}$, k_{ads} = 4 × 10^5 L μg^{-1}min^{-1}, k_{des} = 1 × 10^5 min^{-1}, n = 0.65, and $[S]_T$ = 15,600 μg L^{-1}.

CHAPTER 5
RESULTS AND DISCUSSION OF MODEL SIMULATIONS

Unsaturated (vadose zone) and saturated (groundwater) flow and transport simulations using the numerical models FEMWATER and HGBC123D were used to evaluate the potential field-scale effects of natural chromium sources on ASR operations. Two sets of model simulations were undertaken. The first and more extensive simulations employed a single computational domain to model the unsaturated and saturated zones simultaneously. A second set of simulations was undertaken to test the prospect of decoupling the unsaturated and saturated zone simulations in order to provide a more detailed account of the unsaturated zone while alleviating computational burden. In all simulations, ASR water quality was estimated for local (recovery well) behavior, and global behavior as summarized by spatial moments estimates of the simulated spatial chromium concentration distributions.

SINGLE DOMAIN SIMULATIONS

Scenario I: Mixing of Recharged Water with Chromium in the Background

A uniform initial concentration equal to 3.0e-7 mol/L of Cr(VI) was assumed and the same concentration was assigned to the incoming groundwater flow by means of the variable boundary condition. At the exit boundary, a zero gradient flux boundary condition was used. Recharged water was assumed to contain Cr at 1.0e-11mol/L. Where recharge took place through injection nodes, a constant concentration point source/sink boundary condition was used; where recharge was infiltrated, the variable concentration boundary condition was used at the element faces encompassing the infiltration area (spreading basin).

Sensitivity to Dispersivity

Solutes in groundwater move with the bulk flow of water (advection) and at the same time tend to spread out of this path both in the longitudinal and transversal direction due to mechanical mixing and molecular diffusion (hydrodynamic dispersion). Mechanical mixing or dispersion is caused by the motion of fluid, where some molecules travel faster and others slower than the average flow rate. The parameter that measures this tendency is the dispersivity, and is a characteristic property of the porous medium (Freeze and Cherry, 1979). Molecular diffusion will have an impact on transport only at flow velocities much lower than the ones simulated here, and for this reason its value has been kept constant all throughout the simulations.

The magnitude of the dispersivity is dependent on the transport scale of the problem under consideration. For the case studied here, and based on values reported in the literature (Gelhar et al., 1992), a relevant range of longitudinal dispersivity coefficients extends from 10 m to 110 m. A set of simulations were conducted with dispersivity values in this range, the rest of the parameters being the same as in the base case. The transversal dispersivity (α_T) was always set at 10% of the longitudinal dispersivity (α_L) value. The flow simulation (base case) took 5 minutes and transport simulations took between 8.7 (α_L=10m) and 10.2 minutes (α_L=110m) to complete.

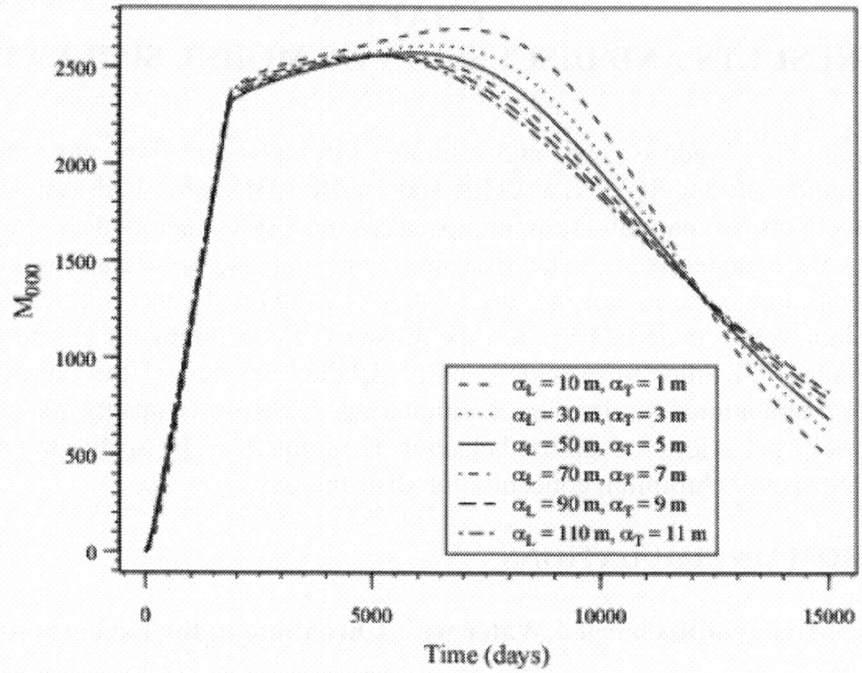

Figure 5.1 The impact of dispersivity on the zeroth moment.

The impact of dispersivity on the zeroth moment (M_{000}) over time is given in Figure 5.1. As expected, lower dispersivity values result in the zeroth moment maintaining a high value for a longer time and decreasing more abruptly than with higher dispersivity: the plume is more compact, it takes longer for its front to reach the exit boundary but when it does it disappears faster from the system. The artificial mass increase mentioned in the methods section is significantly greater for the lower dispersivity values, which are expected to generate steeper concentration fronts. Under these conditions numerical dispersion is created, an inherent problem with solving the partial differential equations governing contaminant transport.

The impact of dispersivity on the trajectory of the center of mass of the clean water plume is given in Figure 5.2. Note that the center of the injection well pattern is at X = 1,410 m, Z = 160 m and Y= 750 m. The Y coordinate of the center of mass remained constant throughout the simulations. After recharge has stopped there is only background flow and only the X coordinate of the center of mass (Xc) is expected to change with time. Figure 5.2 shows that at higher dispersivities the center of mass tends to sink after recharge. Concentration contours for the different dispersivity value simulations suggested an explanation for this: at higher dispersivities the outer boundary of the plume reaches the bottom of the model domain and clean water is accumulated in this area. When the plume starts to leave the model domain the Z coordinate of the center of mass (Zc) starts to increase and Xc changes at a slower rate in Figure 5.2. For smaller dispersivities this happens at a later time and the smaller the dispersivity the closer the center of mass is to the exit boundary at that time because of a more compact plume.

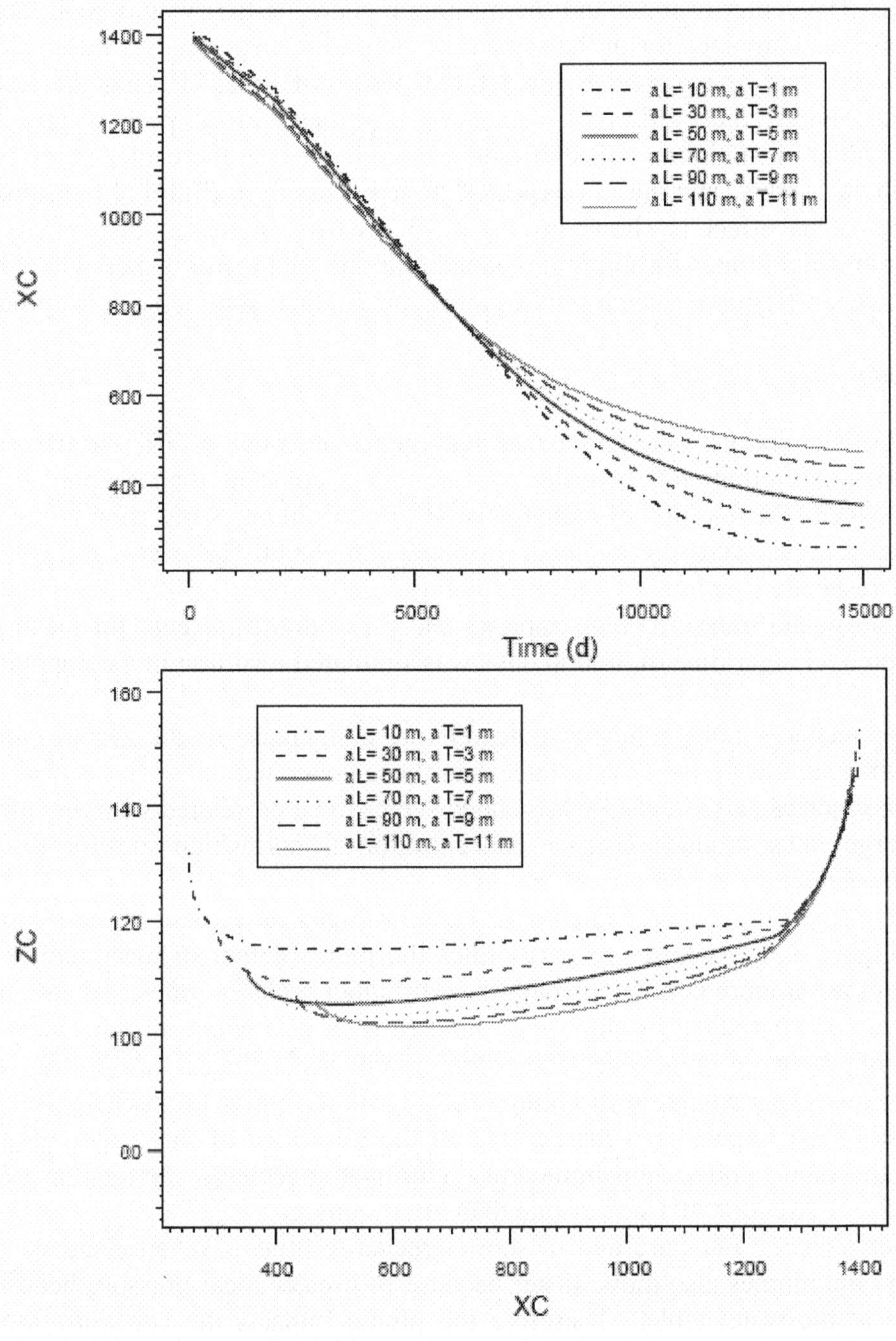

Figure 5.2 The impact of dispersivity on the horizontal displacement (above) and trajectory (below) of the center of mass.

The impact of dispersivity on the average vertical concentration is shown in Figure 5.3 for three locations downgradient of the recharge area. In general, higher dispersivity means a more spread plume, and it is therefore expected that the plume arrives earlier to the observation location, as seen in Figure 5.3. Larger dispersivity causes more mixing with the background water and therefore results in higher values for the minimum concentration. If recharge had been instantaneous, all plumes would be symmetric, with bigger sizes resulting from higher dispersivities. The center of mass and the minimum concentration would be at the center of all plumes and arrive at any location at the same time for all the dispersivity values. Here, however, a constant source has been modeled for a fixed period of time. During the recharge period different dispersivities cause different degrees of asymmetry in the plumes; now the center of mass and the point of minimum concentration are not located in the center of the plume, do not coincide with each other, and are not expected to arrive at any location at the same time for all dispersivities. This effect is shown in Figure 5.3 where increased dispersivity coefficients produce shorter arrival times for minimum concentrations, and lower dispersivity values result in a more symmetric C(t) curve (which implies more longitudinal symmetry in the plume).

Recharge Rate

A set of simulations were conducted injecting clean water at different rates through the 5 recharge nodes configuration. In order to maintain a constant total amount of clean water entering the system, the duration of recharge was varied from one simulation to the next. A total of five recharge rates were investigated: 1,500, 3,000, 6,000 (base case), 9,000, and 12,000 m^3/day with respective periods of injection being 7,200, 3,600, 1,800, 1,200, and 9,000 days. Simulations (flow and transport) took approximately 13 minutes, except for recharge rate=1500 m^3/day in which a longer time of 30,000 days was simulated resulting in a computational time of 25 minutes.

Lower recharge rates result in a slower rate of increase of the zeroth moment (Figure 5.4), and plumes that leave the system later than those generated under fast recharge conditions. Artificial mass increase after recharge has ended seems to be correlated with the rate of recharge: higher recharge rates produce sharper concentration fronts which in turn result in higher numerical dispersion.

The impact of the recharge rate on the center of mass movement of the clean water plume is given in Figure 5.5. At lower recharge rates the plume center of mass moves at shallower depths and follows a more continuous path than at higher recharge rates. At low recharge rates water is moved downgradient by the background flow as it is being introduced into the system whereas at high recharge rates water enters the system faster than it can be transported by the background flow. This results in the plume sinking more during the recharge period at higher recharge rates. The impact of recharge rate on the trajectory of the center of mass tends to decrease as the plume moves downgradient. During the recharge period the center of mass advances in the X direction at a slower rate than after recharge.

Higher recharge rates are tied to shorter recharge times. When clean water is injected more quickly, the plumes also move faster because of greater local pressure head increases due to mounding at the water table. Therefore the plume impacts the concentration at the three observation locations (Figure 5.6) earlier for higher recharge rates than for lower rates. For lower recharge rates the plume generated is vertically thicker and less impacted by mixing with background water. There does not seem to be a straightforward relationship between recharge rate and the value of the minimum concentration. However, this is an outcome of the approach taken here: concentration values are the vertical average throughout the thickness of the aquifer. At low recharge rates the plume moves at shallow depths whereas at high recharge rates the plume tends to sink more. Thus, at low recharge rates the observation points extend outside of the clean water plume and the vertical average used here is less representative of plume behavior. Recharge rate appears to be impact concentrations less at observation locations further downgradient (agreeing with findings for the trajectory of the center of mass).

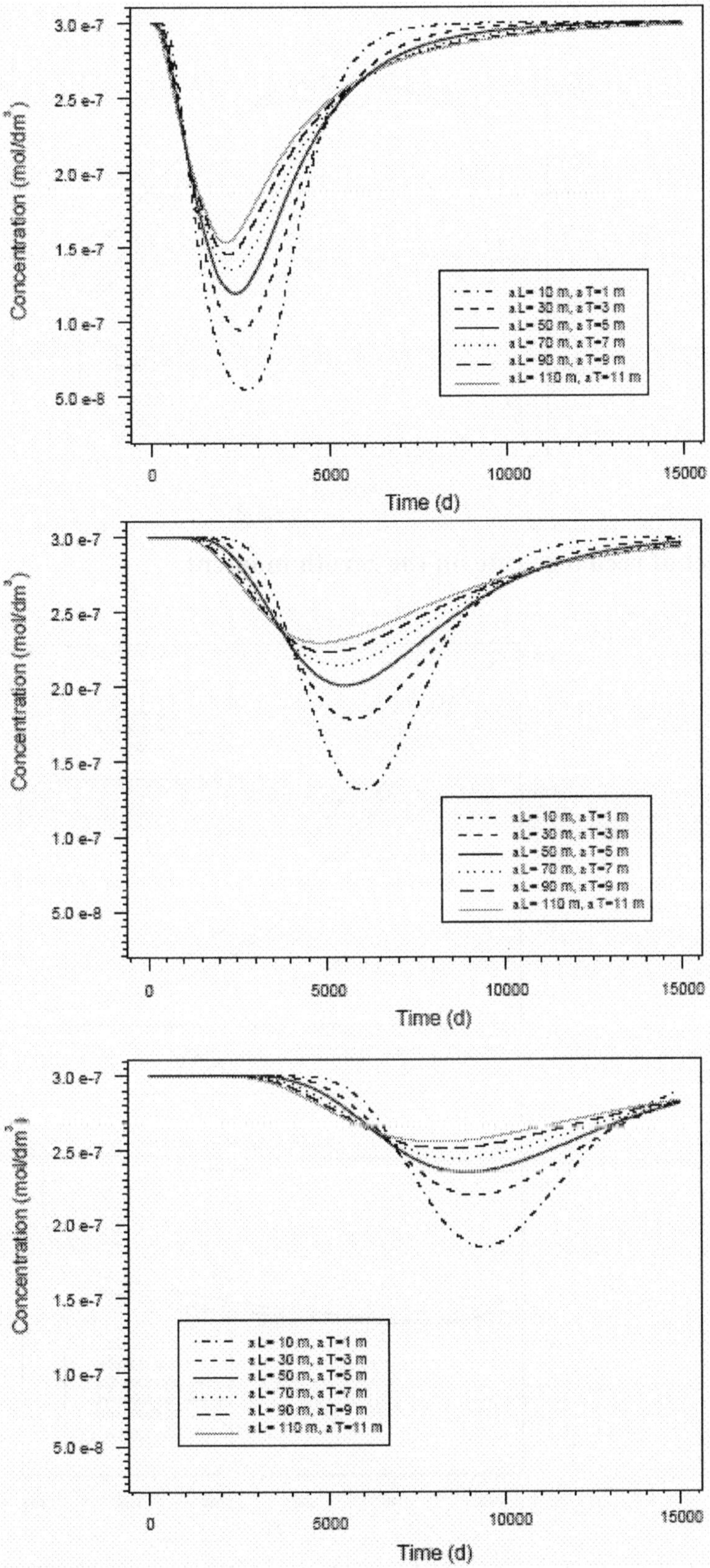

Figure 5.3 Impact of dispersivity on concentration at locations 1 (top), 2 (center), 3 (bottom).

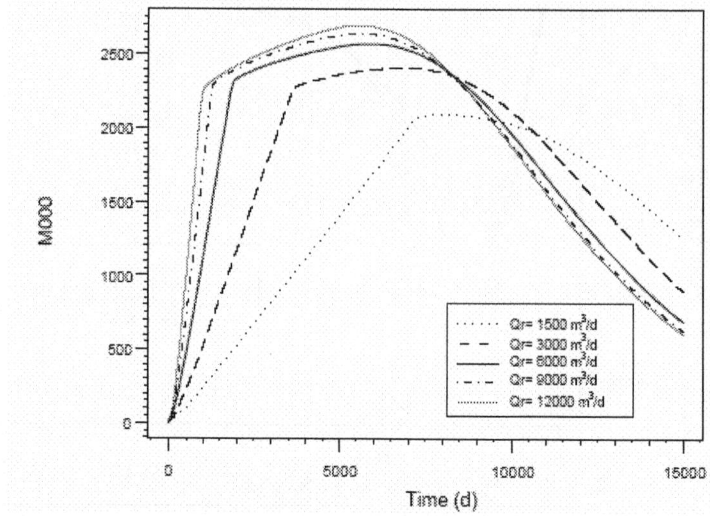

Figure 5.4 Impact of recharge rate on the zeroth moment.

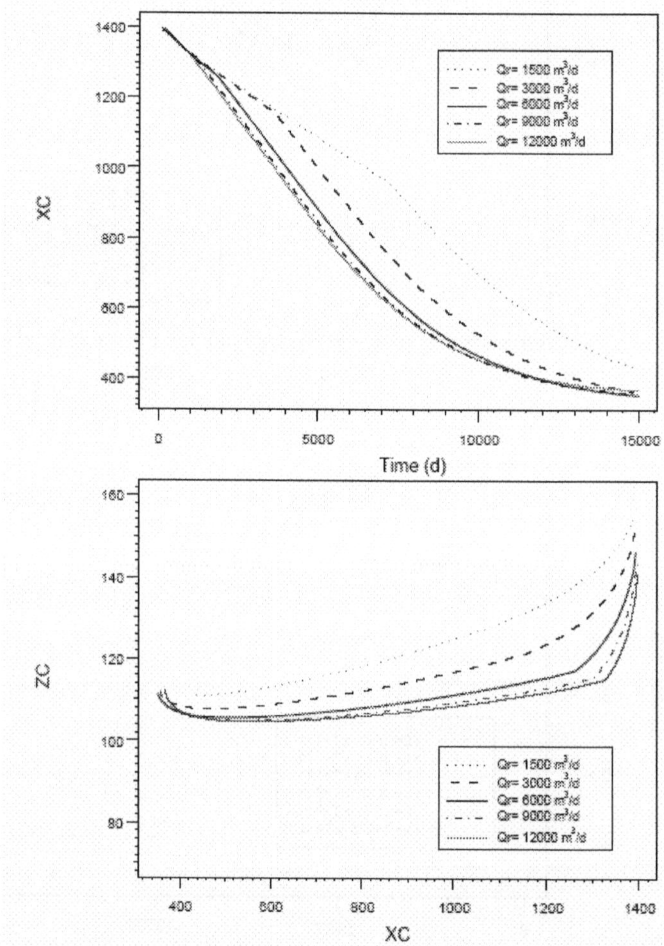

Figure 5.5 Impact of recharge rate on the horizontal displacement (above) and trajectory (below) of the center of mass.

Recharge Configuration

The impact of recharge configuration was studied by simulating recharge through both injection wells and a surface spreading basin. Three cases were simulated where clean water entered the system through injection wells. Each injection well was represented by one recharge node located at an elevation of 160 m (approximately 20 m under the water table). The number and distribution of recharge nodes was varied as shown in Figure 5.7. A total injection rate of 6,000 m^3/day was kept for 1,800 days for each of the configurations with nodal injection rates set at 6,000, 1,200, and 462 m^3/day for the 1, 5 (base case), and 13 recharge node configurations, respectively. A spreading basin covering an area of approximately 20,300 m^2 was also tested, with a recharge rate of approximately 0.3 m/d (Figure 5.8). Although recharging from the ground involves different hydraulic principles than the three injection cases, no preliminary investigation (temporal and spatial discretization) of this case was taken at this point because the goal here was to assess the impact of the recharge configuration only (the rest of the parameters were the same as in the injection wells cases). Total computational time is shown in Table 5-1.

Although the same mass of clean water was introduced in the four configurations, the value of M_{000} is significantly different for all cases (Figure 5.9). Furthermore, there is a pronounced artificial increase for the surface recharge case.

The impact of recharge configuration on the center of mass trajectory of the clean water plume is given in Figure 5.10. For the cases with injection wells, more recharge nodes result in the center of mass moving at shallower depths. This effect is more noticeable close to the wells location, as the plumes move further downgradient the trajectory of the center of mass becomes less sensitive to the configuration of recharge wells. For recharge from a spreading basin, the plume not only starts at a higher Zc position (recharge takes place from the ground surface), but Zc also remains higher throughout the simulation.

The impact of recharge configuration on vertically averaged concentrations is shown in Figure 5.11. More recharge nodes result in a higher minimum concentration (plumes spread horizontally and are vertically thinner, the plume overlaps only some of the observation points included in the average). At the three locations, the concentration time series resulting from surface recharge is similar to concentration time series obtained with the three injection well configurations.

For the system studied here, introducing clean water through injection nodes resembles the model outcome when the same amount of water is infiltrated through a spreading basin. Although there was not an substantial advantage in computational time when modeling recharge through injection nodes, there is a potential for shorter runs and smaller data files by limiting the model domain to the saturated zone with a small portion of unsaturated zone (enough to allow for mounding during recharge when recharge is simulated through injection wells). Whether this simplification is a valid option depends on the degree of accuracy the modeler is willing to compromise. Overall, the ASR simulations are highly sensitive to the recharge configuration in aquifer domains beneath and proximate to the injection or recharge boundary condition. Errors in mass balance can be significant (Figure 5.9) if finer discretization is not employed to more accurately depict groundwater-surface water interactions. The computational burden imposed by the finer discretization is likely to be substantial.

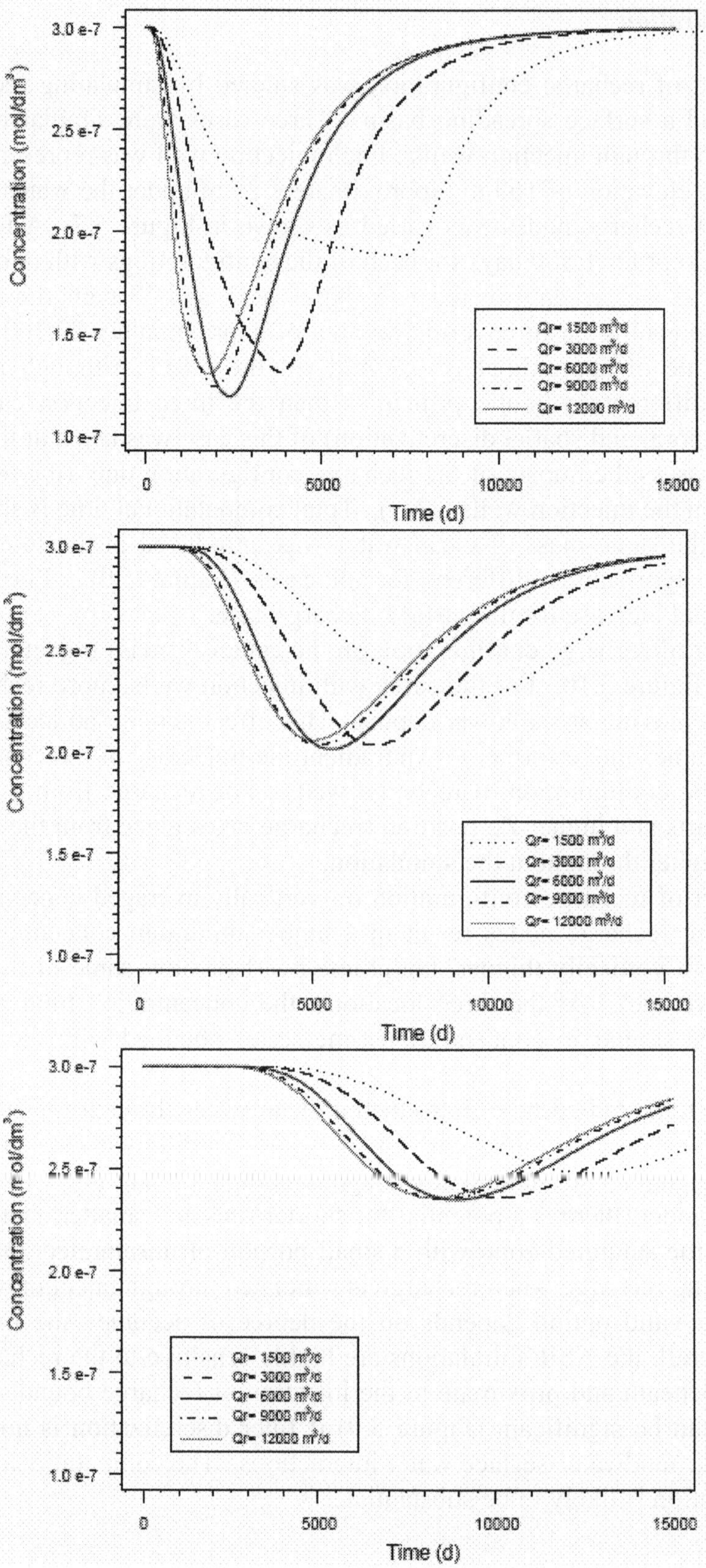

Figure 5.6 Impact of recharge rate on vertically averaged concentrations at three observation locations.

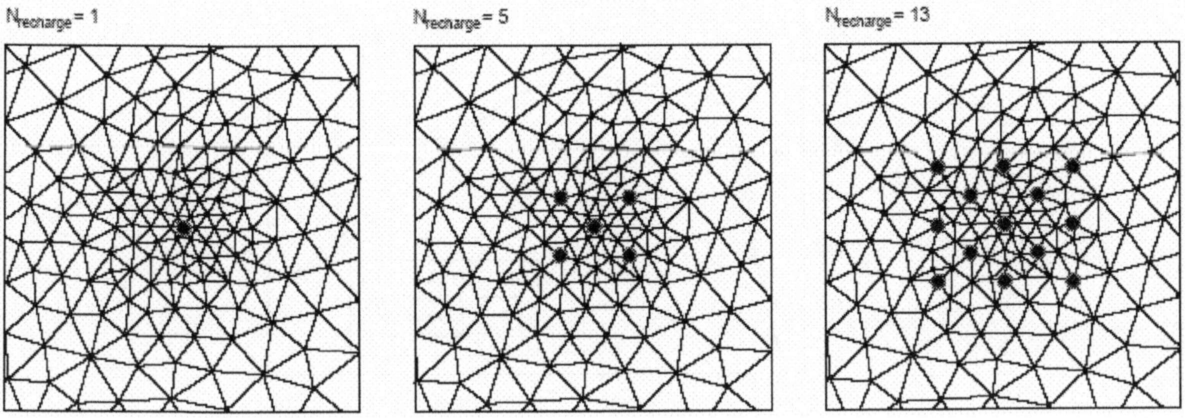

Figure 5.7 Three injection well configurations.

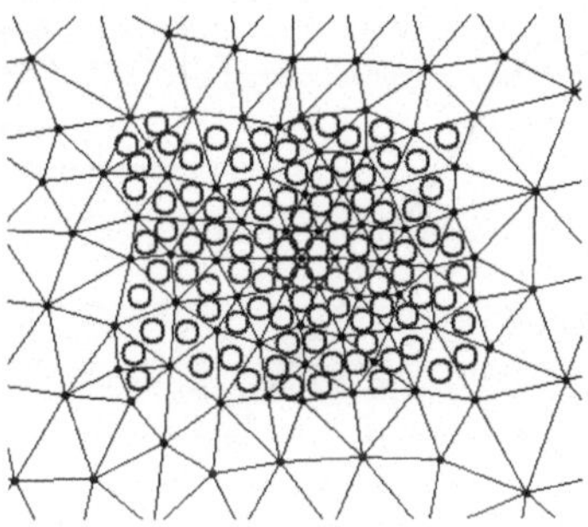

Figure 5.8 Area recharge configuration (spreading basin).

Table 5.1
Computational times for different recharge configurations

Configuration	Flow Run Time (min)	Transport Run Time (min)
one well	3.4	9.6
five wells	5.0	14.8
thirteen wells	3.4	9.4
spreading basin	11.8	10

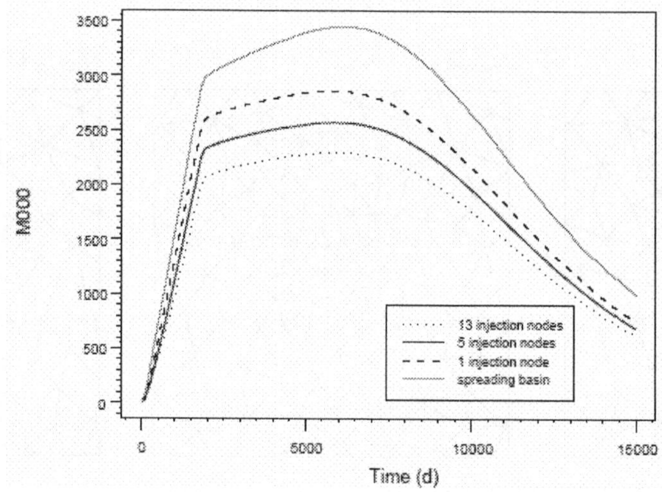

Figure 5.9 Impact of recharge configuration on the zeroth moment.

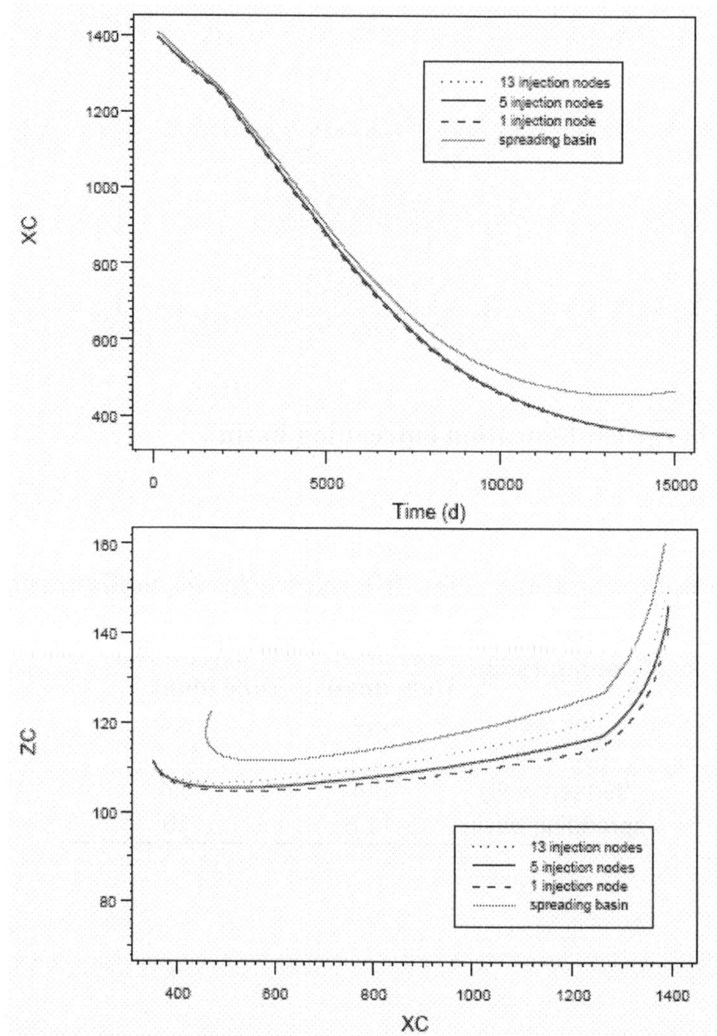

Figure 5.10 Impact of recharge configuration on the horizontal displacement (above) and trajectory (below) of the center of mass.

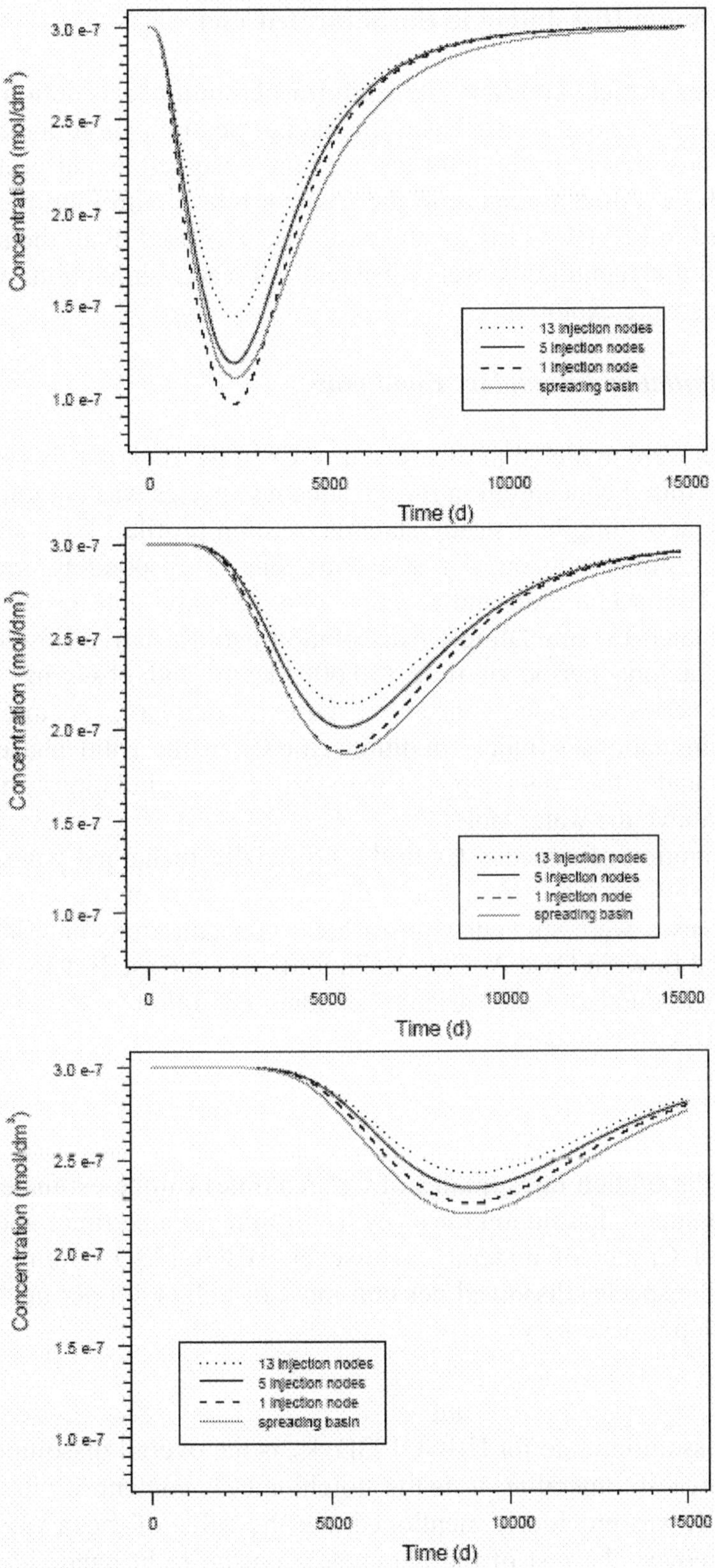

Figure 5.11 Impact of recharge configuration on concentrations at location 200 m (top), 650 m (center), 1100 m (bottom).

Scenario II: Chromium Dissolution in the Saturated Zone

This scenario depicts constant and uniform chromium dissolution as Cr(VI) from the alluvium throughout the groundwater basin, including the simulation domain. In this case, the Cr(VI) concentrations present in the pilot test area are observed as uniform because of the long travel time through the domain relative to the weather rates controlling the Cr(VI) release. The flow field generated in FEMWATER for the base case was used in all the simulations presented in this section. Initial conditions were different for each simulation, whereas two sets of boundary conditions were explored.

Transport Model Initial and Boundary Conditions

Two species were modeled in this scenario: Cr in the rock matrix (solid phase) and Cr in the aqueous phase. Initial concentration for the aqueous species was not uniform as in Scenario I but varied spatially according to a steady state dissolution profile. Each simulation contained a different pattern of initial aqueous Cr concentrations corresponding incoming groundwater (native background) concentration and reaction rate. Steady state groundwater concentration profiles were calculated by simulating Cr dissolution against native flow (no recharge, natural gradient only) for a long period of time (50,000 days), and observing at a time such that concentration time series remained constant at the exit boundary. A time of 50,000 days was sufficient for all simulations. Initial conditions for Cr in the solid phase were set so excess precipitate existed under the water table at the start of the simulations and no Cr in the solid phase was present above the water table.

The concentration of aqueous Cr in the artificially recharged water was assumed to be negligible (1e-11 mol/L). The front, back, top, and bottom of the box were set as no flux boundaries. Two cases were studied with different concentrations entering the model domain with the background flow: in Case A, Cin=3e-7 mol/L and in Case B, Cin=1e-11 mol/L. Both at the entrance and exit boundaries the variable boundary condition type was used. All simulations use C=0 mol/L at the exit boundary.

Dissolution Rates

Chromium dissolution rates (moles of Cr/area/time) can be estimated based on literature values for silicate mineral dissolution rates (moles of mineral/area/time) and Cr concentration in the mineral (mol of Cr/mol of mineral). However, HBGC123D requires dissolution rates in terms of moles of the species dissolved per unit mass of bulk solid per unit time. Therefore the following relationship is necessary:

$$k_{HBGC} = \frac{A_m}{m_s} v_i k_m \tag{5-1}$$

where k_{HBGC} is the dissolution rate for HBGC123D, k_m is the overall dissolution rate of mineral *m*, A_m is the surface area of mineral *m*, v_i is the stoichiometric content of *i* in mineral *m*, *and* m_s is the mass of solid. It was previously mentioned that the value of *k* can vary over several orders of magnitude. Moreover, the rest of the parameters on the right hand side of the equation are very specific to the type of minerals present in the aquifer and their abundance. Thus, it is not possible to establish a range based on literature values. For this reason a different approach was followed to choose a range of dissolution rate constants. It was assumed that the Cr concentration observed in the Fenner Gap area during the pilot tests (approximately 3e-7 mol/L) is the result of continuous dissolution of Cr from aquifer materials along the natural recharge-groundwater flow path. Assuming naturally recharged water follows a linear path between the New York mountains and Fenner Gap (a distance of approximately 50 km), that the average pore water velocity in the watershed is the same as in Fenner Gap (approximately 0.3 m/d), and that soil properties are uniform throughout the watershed and the same as those in Fenner Gap, a rate constant of approximately 5e-13 mol/kg of solid/day was obtained. All of these assumptions are approximate. However, they are a starting point for choosing a reaction rate. A range of zero order kinetics reaction rates from 1e-13 to 1e-8 mol/kg of solid/day was simulated.

Transient simulations (50,000 days) to achieve steady state concentration distributions took approximately 45 min, whereas transport simulations with the base case flow (15,000 days) took approximately 15 min.

Case A. The plot in Figure 5.12 depicts concentration normalized by the initial concentration for the different reaction rates simulated under a background concentration of 3e-7 mol/L. For k values of 1e-12 mol/kg/d and lower this value closely resembles that for the non-reactive transport base case. At reaction rates of 1e-9 mol/kg/d and above normalized concentrations seem to be insensitive to the reaction rate. Because initial concentrations are different for each case, the same cannot be said about absolute concentrations. Figure 5.13 shows concentration time series for two different reaction rates.

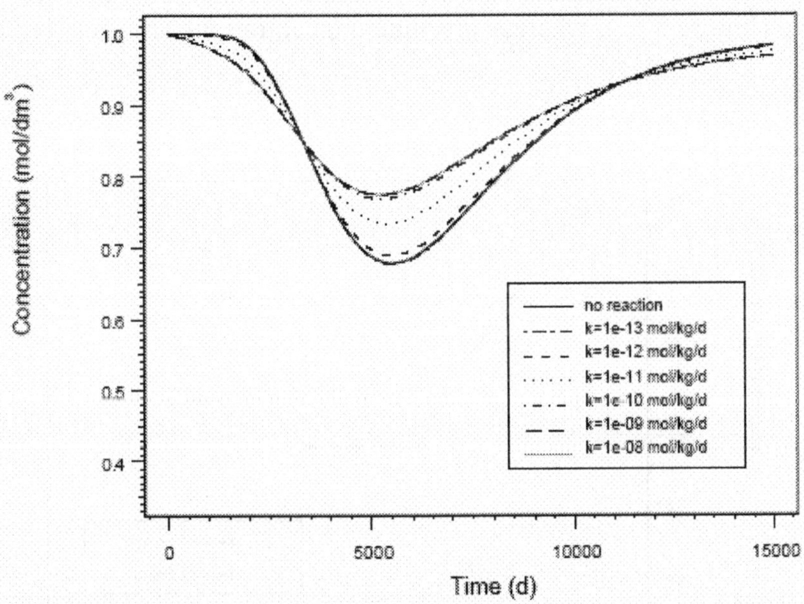

Figure 5.12 Impact of dissolution rate on normalized concentration at observation point 650 m downgradient from the recharge area.

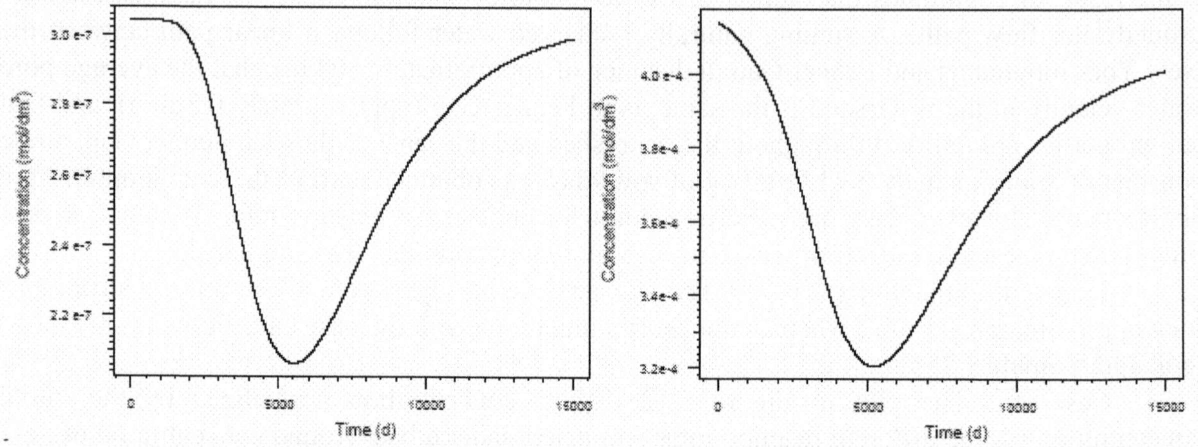

Figure 5.13 Concentration time series (650 m location) for k=1e-13 mol/kg/day (left) and k=1e-8 mol/kg/day (right).

Case B. A base case restricted to mixing between the artificially recharged water and the groundwater at background concentrations could not be established because groundwater background and artificially introduced concentrations were the same. Under an incoming concentration of 1e-11 mol/L, normalized concentrations are insensitive to the values of k in the range studied here (Figure 5.14), whereas absolute concentrations are sensitive to this parameter (Figure 5.15).

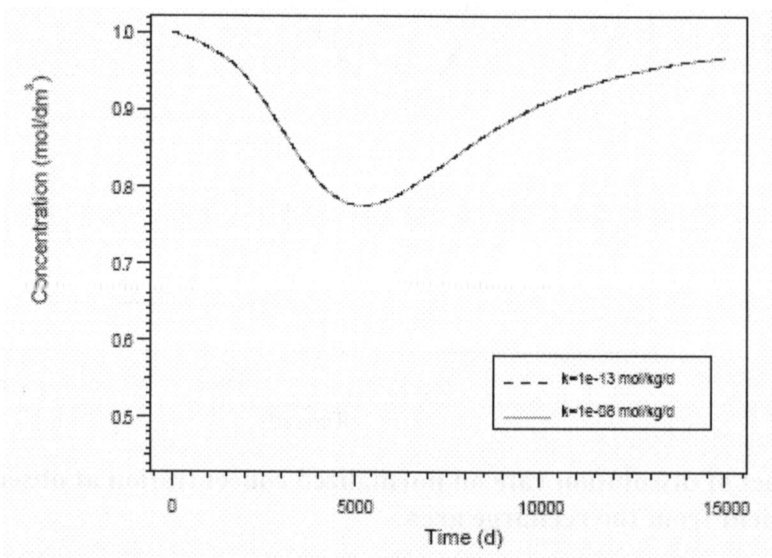

Figure 5.14 Impact of dissolution rate on normalized concentration 650 m downgradient from the recharge area.

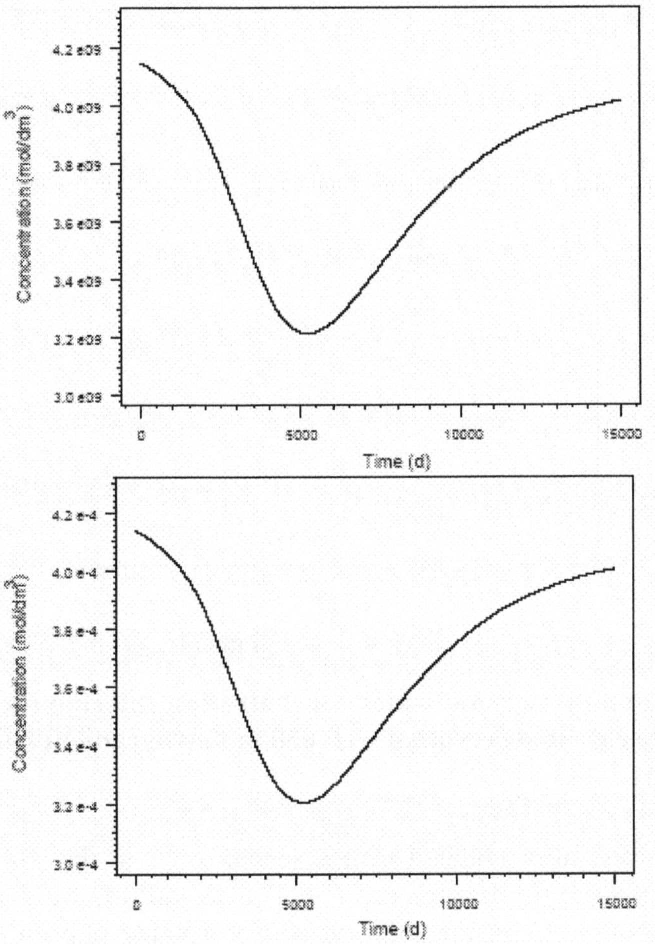

Figure 5.15 Concentration time series (650 m location) for k=1e-13 mol/kg/day (left) and k=1e-8 mol/kg/day (right).

Case A vs Case B. For k=1e-13 mol/kg/day, both normalized (Figure 5.16) and absolute concentrations (Figures 5.13 and 5.15, left) depend on the concentration entering the model domain with the groundwater flow. For k=1e-8 mol/kg/day, concentrations become independent of the native groundwater concentration entering the model domain.

Scenario III: Chromium Dissolution in the Unsaturated Zone

Reactive transport simulations were carried with the code HBGC123D. A flow field similar to that used in the infiltration case in Scenario I was used, however, HBGC required a finer time resolution during the recharge stage to converge. To understand the impact of Δt in HBGC, the same case of spreading basin infiltration as in Scenario I (clean water against a background with C=3.0e-7 mol/L) was modeled. Time increments of 100, 50 and 5 days were tested, with HBGC being able to run only at Δt =5 days. Thus, time steps were set at 5 days for the recharge period and at 100 days after 2,000 days (when all recharge mounding effects have disappeared).

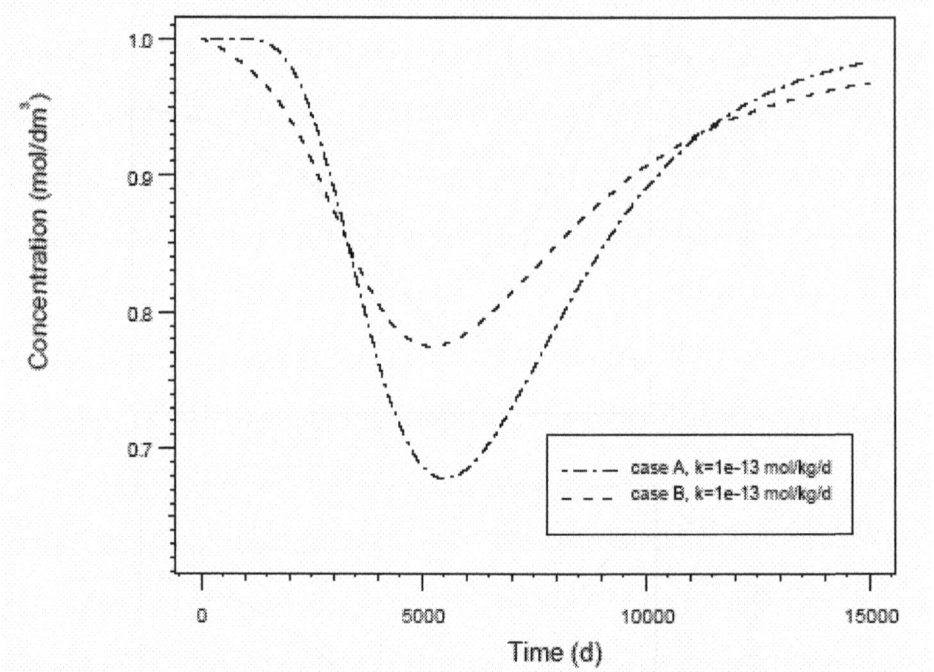

Figure 5.16 Impact of native groundwater concentration entering the simulation domain on concentration time series at observation well 650 m downgradient from recharge area (k=1e-13 mol/kg/day).

Two species were also modeled in this scenario: Cr in the soil and Cr in the aqueous phase. Initial concentration for the aqueous species was uniform throughout the model domain and equal to 1e-11 mol/L. Cr in the solid phase was located only in the zone underlying the spreading basin above the water table. The initial amount of Cr in the solid phase was established such that there was sufficient Cr to last for the duration of the recharge period.

Concentration of aqueous Cr in the infiltrated water was assumed to be negligible (1e-11 mol/L) through the variable boundary condition. The front, back top and bottom of the box were set as no flux boundaries. Concentration entering the model domain was set at 1e-11 mol/L. All simulations use C=0 mol/L at the exit boundary. Both at the entrance and exit boundaries the variable boundary condition type was used. Figure 5.17 shows the zeroth moment for k=1e-13 mol/kg solid/d and k=1e-8 mol/kg solid/d. Dissolution only takes place during the recharge period, therefore Mooo only increases until day 1800. The behavior of the zeroth moment is similar to that observed under Scenario I. In Scenario II, the model contained a constant source of clean water, in this case the model used a constant source of the contaminant (consequence of zero order kinetics of dissolution). As expected, for each order of magnitude increase in the value of k there is an order of magnitude increase in the mass introduced in the system at the end of the recharge-dissolution period.

The pattern of concentration time series (Figure 5.18) is inverted with respect to those in Scenarios I and II because Cr initial concentrations were negligible all throughout the saturated zone, and dissolution took place only in the recharged water. There is no upper bound for the normalized (and absolute concentration) as the value of k increases, because there was no reaction in the background water.

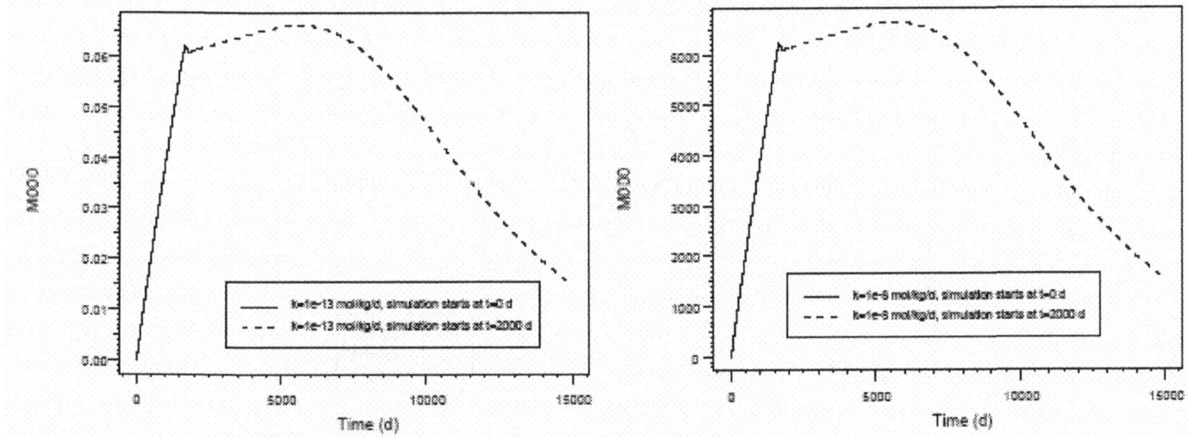

Figure 5.17 Impact of dissolution rate on the zeroth moment.

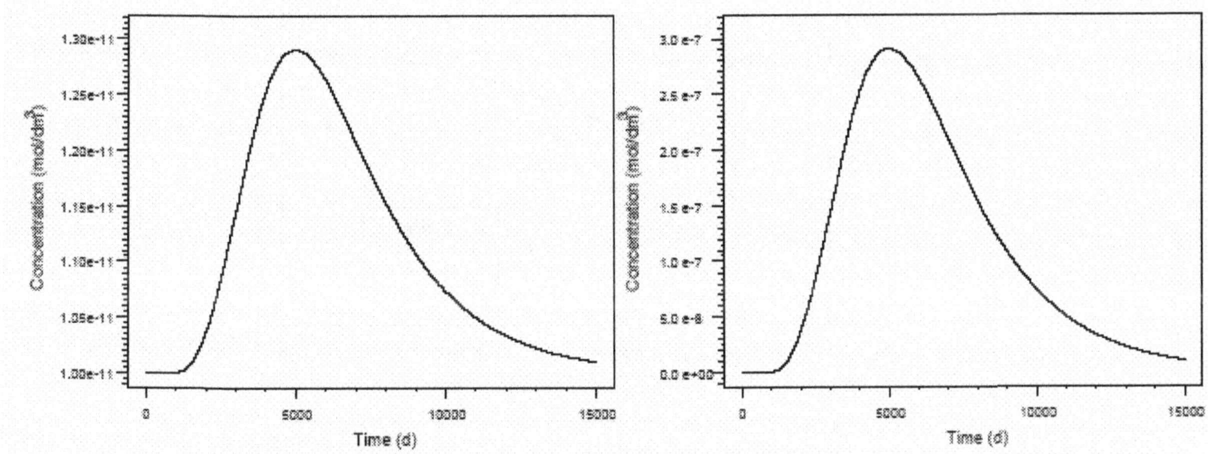

Figure 5.18 Concentration time series at observation location 2 for k=1e-13 mol/kg/day (left) and k=1e-8 mol/kg/day (right).

Comparison of Source Scenarios I, II, and III

The temporal variation of the quality of recovered water from an ASR system will be markedly different depending on the location and type of chromium source (Figure 5.19). When clean water is injected against a chromium-containing background (Scenario I), the quality of the injected water diminishes (minimum concentration is higher) as result of mixing the further the plume moves downgradient. When dissolution takes place in the saturated zone (Scenario II), recharged water quality degrades faster, the result of both mixing and dissolution. When dissolution takes place in the unsaturated zone against a clean background, further downgradient displacement of the recharged water helps to improve the quality of the plume by dilution with native water.

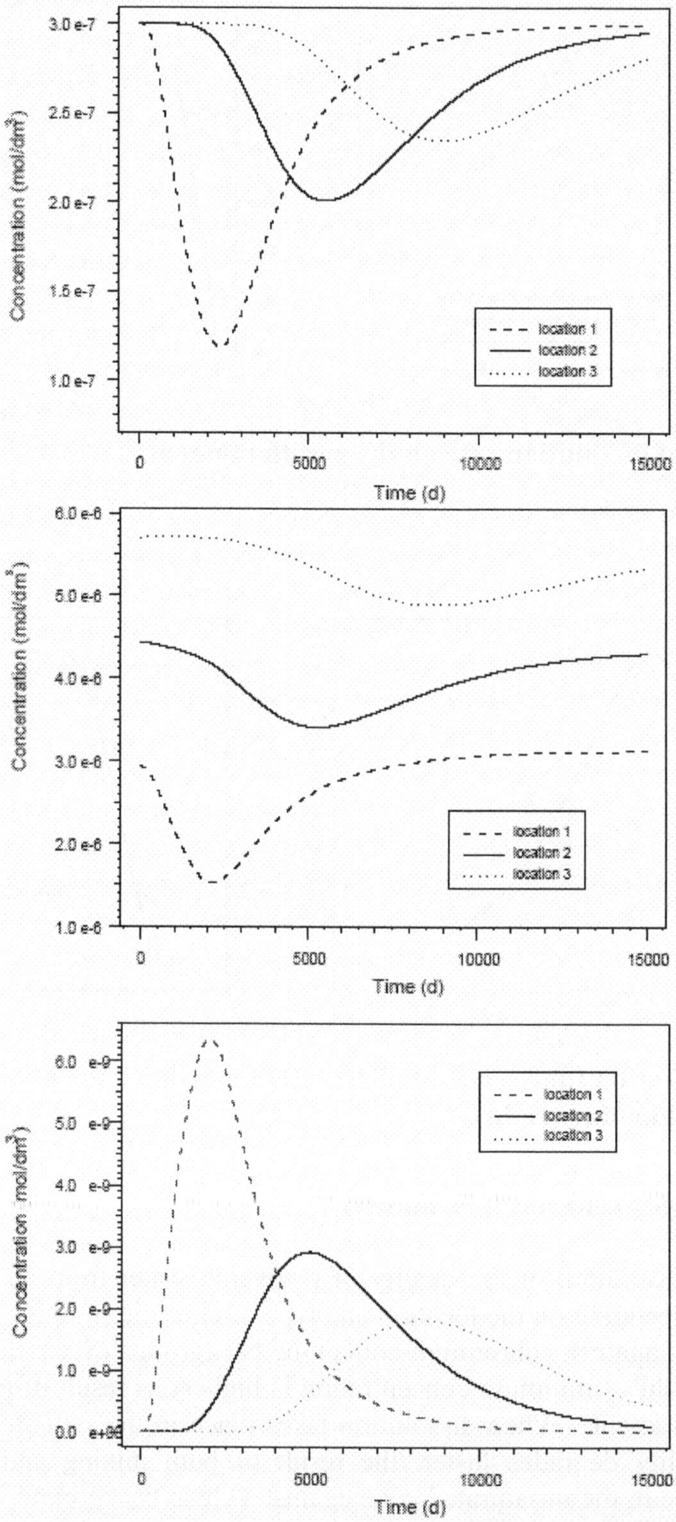

Figure 5.19 Concentration time series at locations 1 (200 m), 2 (650 m) and 3 (1100 m). Scenario I: recharged water mixing with Cr in background water (top), Scenario II: dissolution in the saturated zone (center), k=1e-10 mol/kg/day, Cin=3e-7 mol/L, Scenario III: dissolution in the unsaturated zone, k=1e-10 mol/kg/day.

DUAL DOMAIN SIMULATION RESULTS

The computational requirements for modeling flow and transport under unsaturated conditions are much greater than those required for saturated conditions. In the dual domain approach, the conjunctive use system is separated into two separate models (System 1 and System 2). System 1 models the unsaturated zone from the spreading basin (350 m above datum) to the aquifers water table (183 m above datum). System 2 models the unconfined aquifer from bedrock (0 m above datum) to approximately 7 m above the water table (190 m above datum) with the bulk of the unsaturated zone eliminated from the models spatial domain.

System 1 Simulations

Output from the System 1 flow and transport simulation gives the nodal pressure heads and concentrations for each time step in the simulation. The plots in Figure 5.20 depict the Cr concentration distributions within a vertical cross section (Y = 750 m) of the System 1 box after 10, 20.5, 40, and 60 days of simulation. At 10 days, the concentration gradient is observed to increase with depth as the Cr associated with the soil moisture is swept downward by the infiltration clean water. However, as time progresses (see 20.5 day contours), this gradient has disappeared due to the absence of residual aqueous Cr in the flowpath. A steady-state distribution is achieved at some time between 20.5 and 40 days.

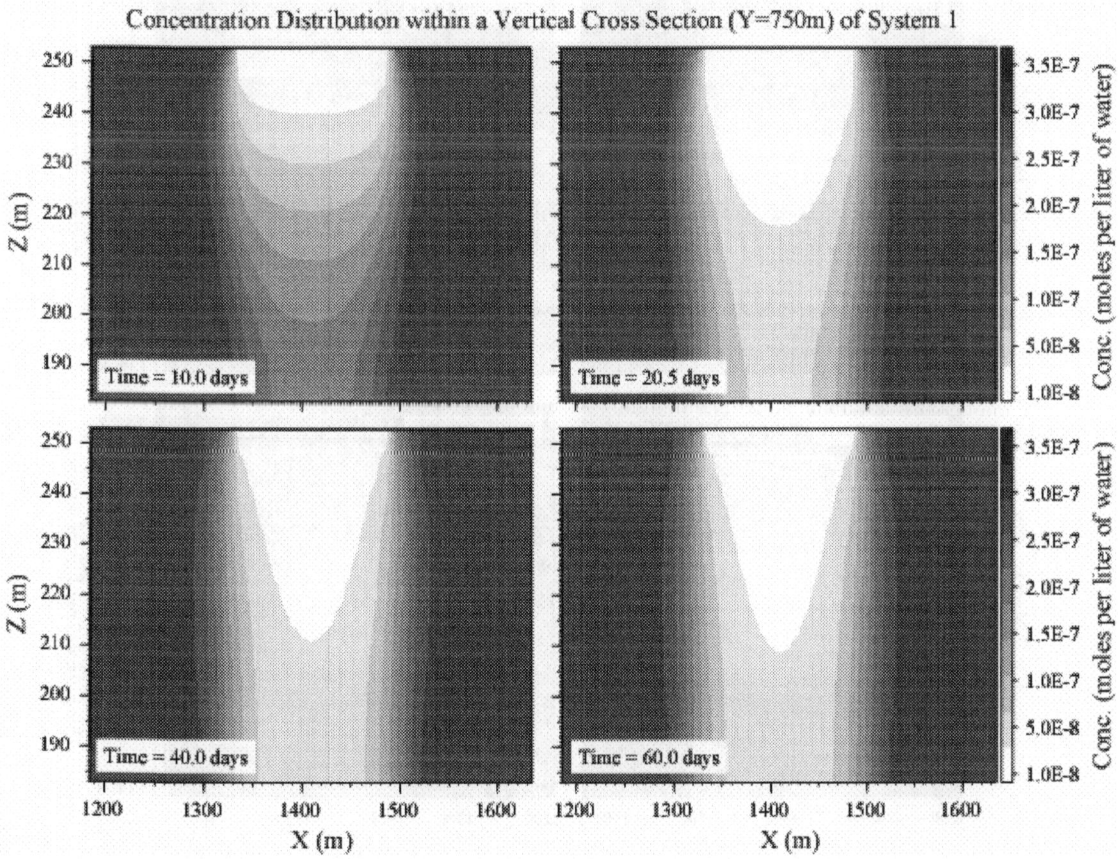

Figure 5.20 System 1 Cr concentration distributions within a vertical cross-section.

The plots in Figure 5.21 depict the Cr concentration distributions within a horizontal cross section (Z = 190 m) of the System 1 box after 10, 20.5, 40, and 60 days of simulation. For the same reasons as were discussed for the vertical contours, the Cr concentrations are greater at the earlier time, but rapidly decrease as a steady, Cr-free flow path develops between 20.5 and 40 days.

Recharge Boundary Condition

Implementation of the Recharge Boundary Construction Algorithm using simulation results from System 1 generates the recharge boundary conditions for System 2. Output from the algorithm gives the fluid flux and Cr concentration for each element face in the System 2 recharge boundary. Figure 5.22 describes the System 2 recharge flow boundary after 10 and 40 days of simulation. The fluid flux distribution after 10 days is characterized by high flows with little spreading. Recall that the recharge flow boundary of System 2 describes the System 1 flows passing through a cutting plane at Z = 190 m. Therefore, after 10 days of simulation the recharge flow boundary of System 2 would describe the upper portion of the conical wetting front as it expands towards the water table. The low flows and high spread after 40 days of simulation describe a period of drainage after recharge has ended.

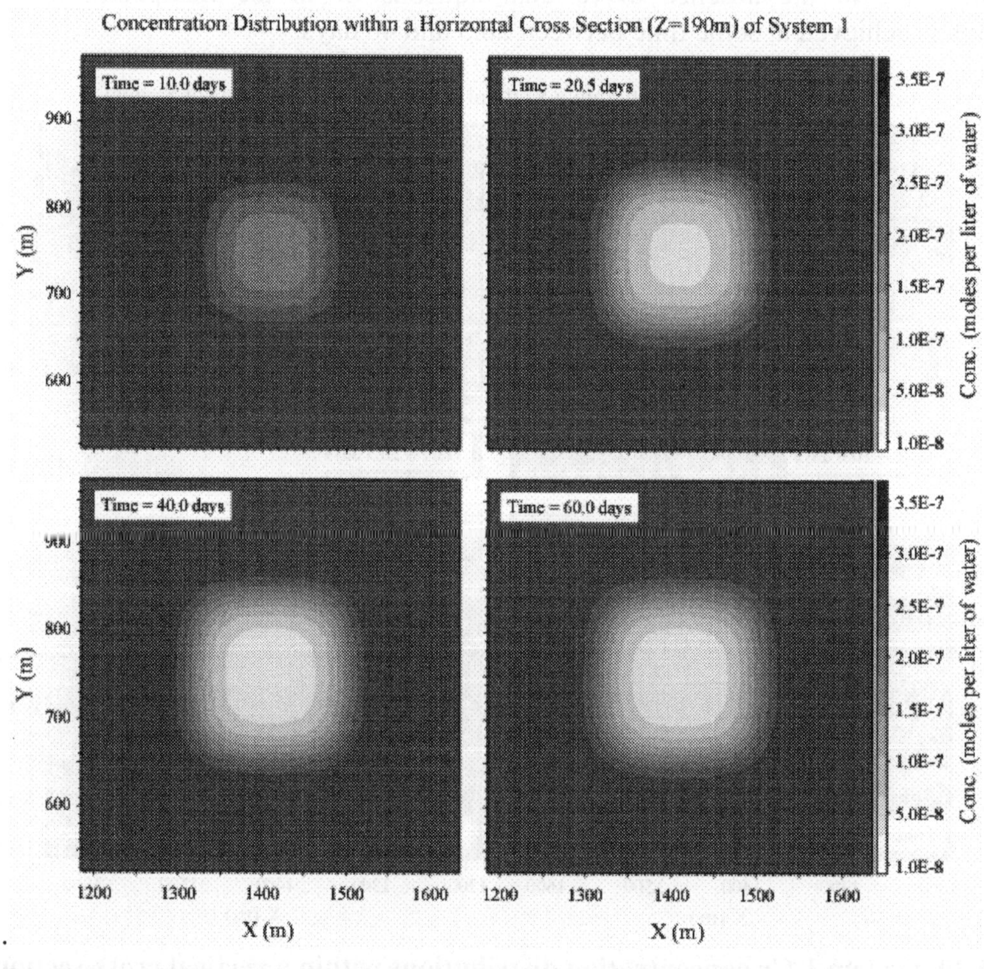

Figure 5.21 System 1 Cr concentration distributions within a horizontal cross-section.

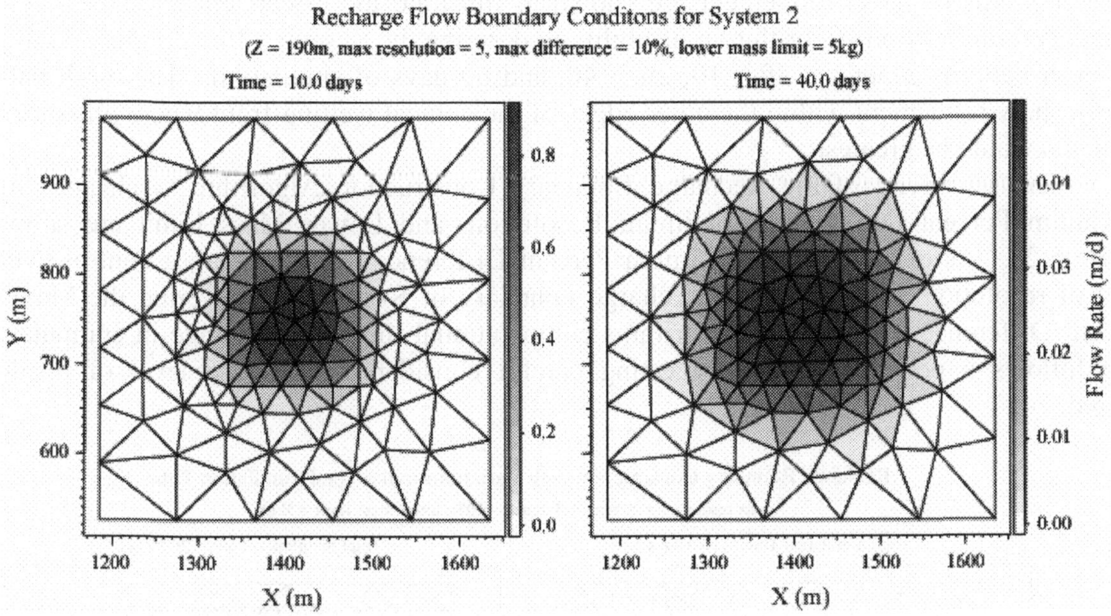

Figure 5.22 Recharge flow boundary conditions for System 2.

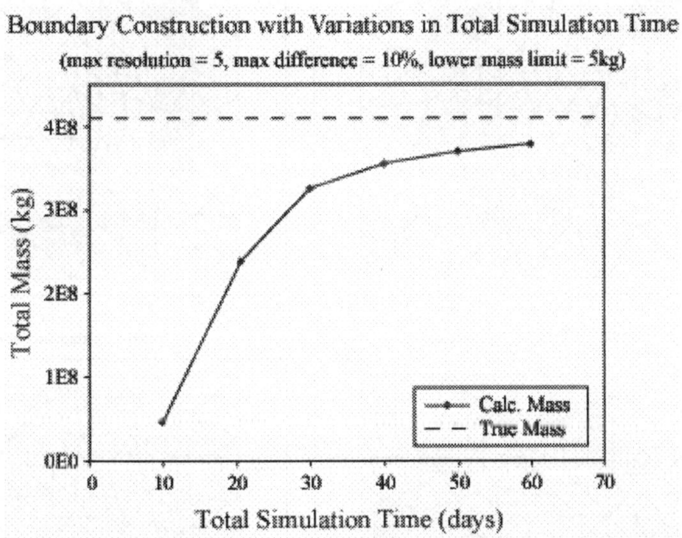

Figure 5.23 Sensitivity of the Recharge Boundary Construction Algorithm to variations in the total simulation time.

The Total Mass is defined as the total fluid mass passing through the System 2 recharge boundary during a System 1 simulation. By varying the total simulation time of System 1 the Total Mass may be calculated for any length of time. Figure 5.23 gives the Total Mass at varying times in the simulation. The greatest increase in Total Mass is observed going from day 10 to day 20. Recall that the recharge from the spreading basin ends after 20.5 days and is followed by a period of drainage, thus explaining the decrease in the change in Total Mass rate over time. The longer the system is allowed to drain the closer the Total Mass comes to the exact mass of recharged water entering through the spreading basin (409,199,928 kg).

Patterns of mesh refinement are shown to illustrate the effectiveness of the Recharge Boundary Construction Algorithm in its ability to describe the System 1 flow. Figure 5.24 gives the mesh patterns generated after 10, 20.5, 40, and 60 days of simulation. The mesh patterns clearly show refinement within the outer edges of the conical wetting front where pressure head gradients would be greatest.

The sensitivity of the Recharge Boundary Construction Algorithm to changes in the Maximum Percent Difference, Maximum Resolution, and Lower Mass Limit are shown in Figure 5.25. As expected, the Maximum Percent Difference has the greatest control over the level of resolution described in the recharge boundary of System 2. Lowering the Maximum Percent Difference increases the algorithms computational time. The change in computational time follows a power function suggesting careful consideration for the level of resolution desired.

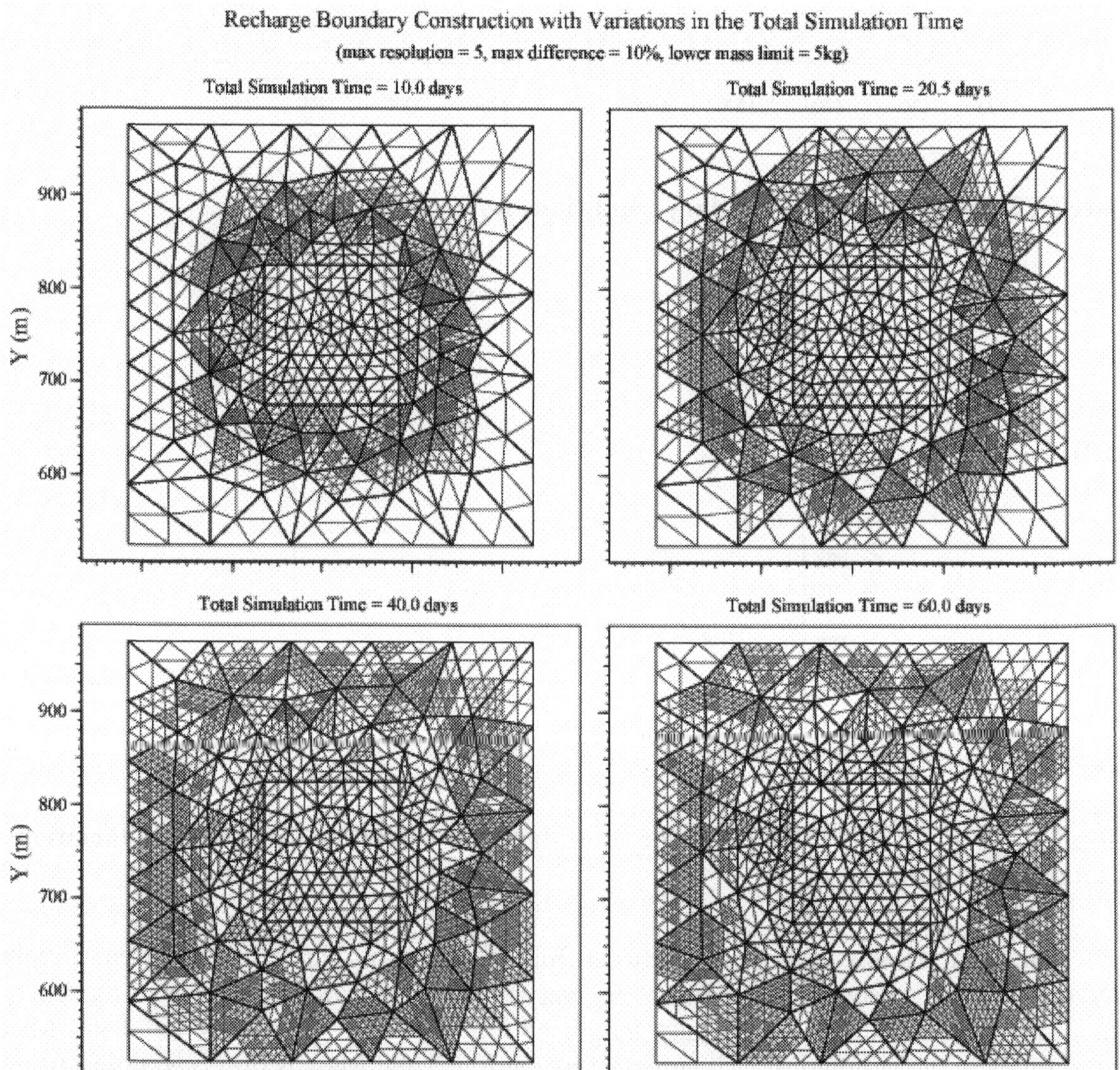

Figure 5.24 Mesh development with the Recharge Boundary Construction Algorithm.

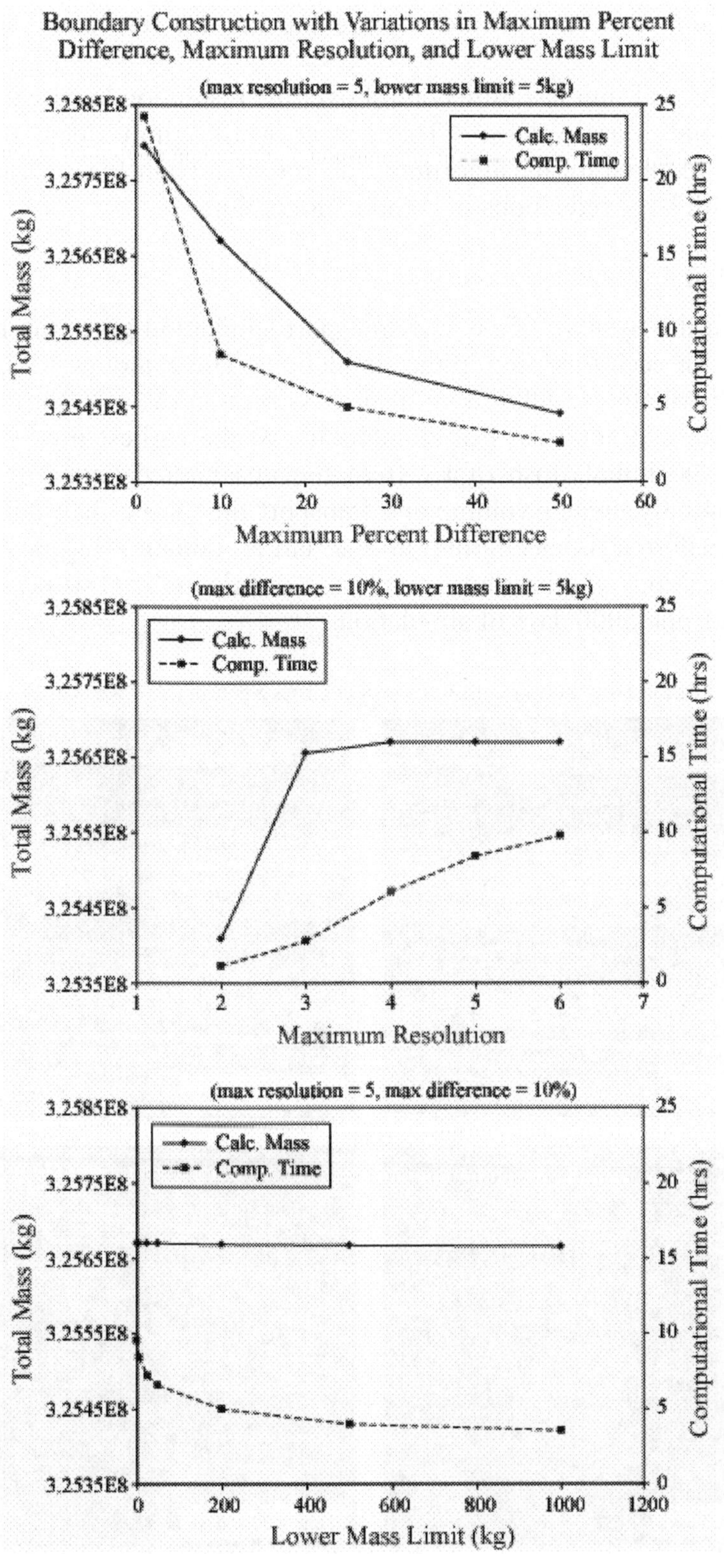

Figure 5.25 Sensitivity of the Recharge Boundary Construction Algorithm to variations in the Maximum Resolution, Maximum Percent Difference, and Lower Mass Limit.

For Maximum Resolutions greater than 3 there is little gained by enhanced resolution. A Maximum Resolution of 3 or 4 would capture the highest resolution at the lowest cost in computational time.

The Lower Mass Limit has little control over the resolution level with minimal resolution loss for higher Total Mass values. The Lower Mass Limit does, however, impact the computational time required for the algorithm. Therefore, a High Mass Limit will result in small resolution losses with large reductions in computational time.

System 2 Simulation

Output from the System 2 flow and transport simulation gives the nodal pressure heads and concentrations for each time step in the simulation. The contours in Figure 5.26 exhibit the Cr concentration distributions within a vertical cross section (Y = 750 m) of the System 2 box after 50, 1000, 3000, and 5000 days of simulation. At the earliest time (50 days), there is no apparent effect of the initial flush of Cr from the unsaturated zone (System 1). System 2 represents a substantially greater volume and transport time scale than does System 1, and the amount of Cr transferred is negligible in the larger domain. Figure 5.27 shows the Cr concentration distributions within a horizontal cross section (Z = 180 m) of the System 1 box after 50, 1000, 3000, and 5000 days of simulation.

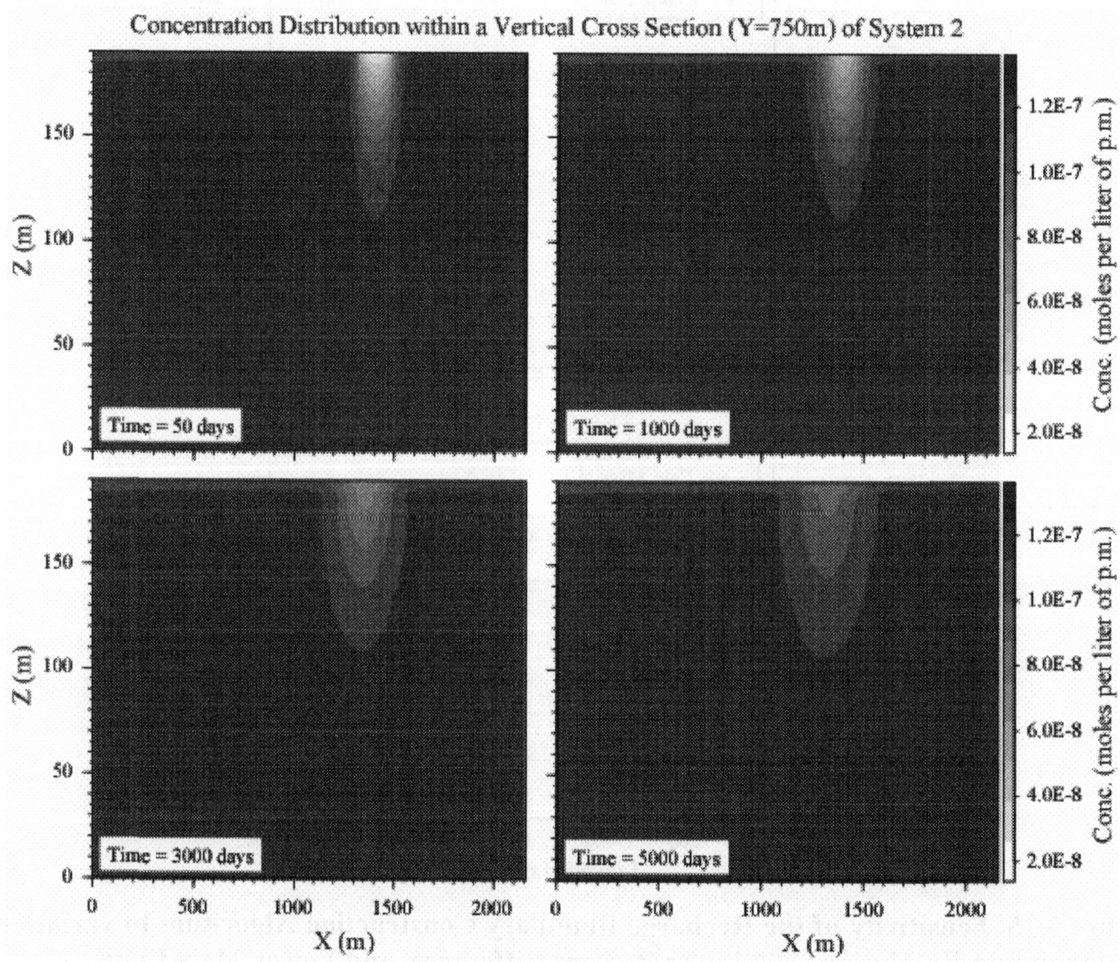

Figure 5.26 System 2 Cr concentration distributions within a vertical cross-section.

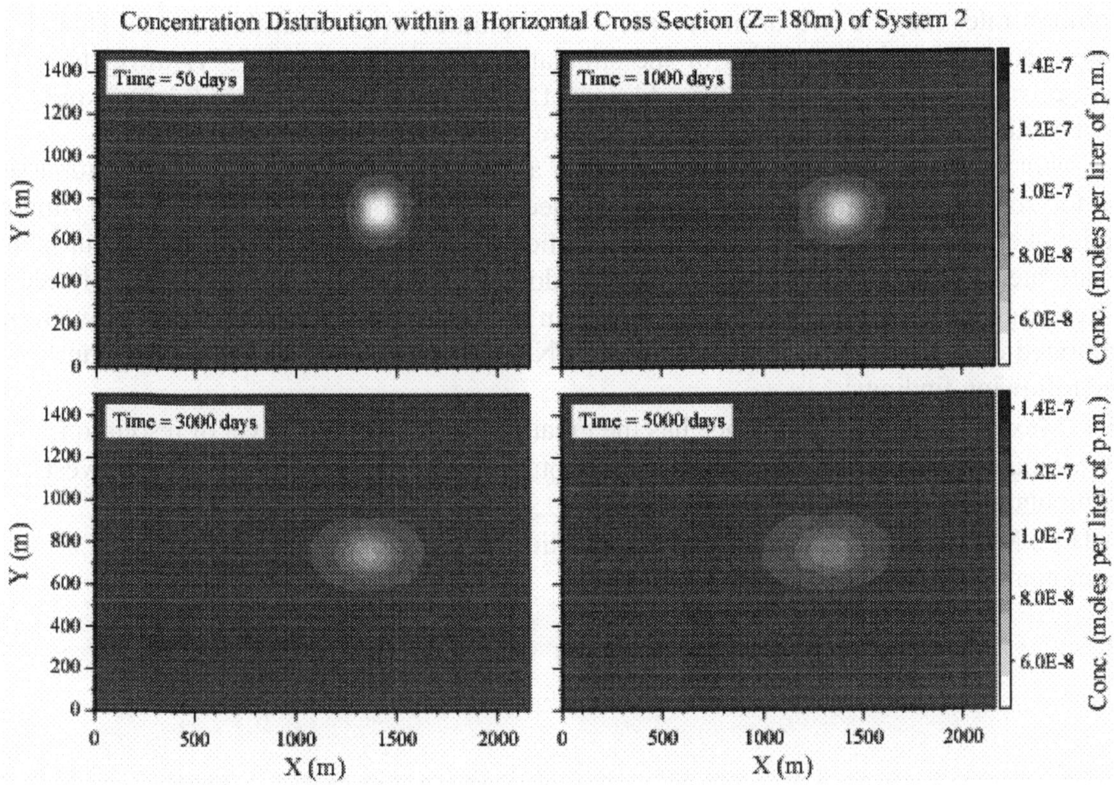

Figure 5.27 System 2 Cr concentration distributions within a horizontal cross-section.

SUMMARY AND CONCLUSIONS

Model simulations of the ASR system were first undertaken using a single-domain model that was computationally demanding but demonstrated the effects of key parameters on the water recovery process. The greatest impact on the zeroth moment (mass of clean water recovered) resulted from changes in recharge rate and recharge configuration. The horizontal movement and trajectory of the clean water plume was most affected by recharge rate (and time length of the recharge period). Concentration time series at observation points downgradient from the recharge zone were most sensitive to the value of dispersivity.

For the domain and conditions simulated, the impact of dissolution rate on normalized local concentrations downgradient are affected by the value of incoming concentration. Incoming concentration had a significant effect at the low reaction rates, but the effect lessened with increasing reaction rate. There was an upper bound for the normalized concentration when dissolution into the recharged water took place against a background that had reached steady state. Cr weathering/dissolution rates of 1e-12 mol/kg/d and lower were sufficiently slow as to produce output that closely resembled that for the non-reactive transport base. This means that the dissolution rate could be safely neglected in these cases. For reaction rates of 1e-9 mol/kg/d and greater, local equilibrium was valid throughout the simulated domain. Intermediate rates (1e-12 mol/kg/d < k < 1e-9 mol/kg/d) produced significant effects on the recovered water quality for the conditions simulated. The results from the literature suggest that silicate weathering rates are extremely low, ranging from 10^{-17} to 10^{-11} mol/m^{2}s. Given that the stoichiometric content of chromium in most silicate minerals is expected to relatively low (< 1%), congruent chromium dissolution rates may be two or more orders of magnitude lower than this range. Thus, for the domain geometry and time scales of the simulations shown here (10-15 yr travel times from recharge zone to recovery wells), the addition of Cr to artificially recharged water would occur almost exclusively through mixing with the native groundwater.

An innovative numerical approach was also presented here in which unsaturated and saturated zone flow and transport process were decoupled. First, the unsaturated zone simulation is undertaken in a significantly smaller computational grid than is required when the unsaturated and saturated media are combined in the single-domain approach. Then, a method referred to as the Recharge Boundary Construction Algorithm was employed to transfer the flow from the unsaturated domain into the saturated domain. Numerical testing of the new algorithm resulted in the following findings:

- The Maximum Percent Difference has the greatest control over the level of resolution.
- A Maximum Resolution of 3 or 4 would capture the highest resolution at the lowest cost in computational time.
- A High Mass Limit will result in small resolution losses with large reductions in computational time.

CHAPTER 6
SUMMARY AND CONCLUSIONS

The processes by which Cr is partitioned between solid or immobile and dissolved or mobile phases are critical both to the occurrence and mobility of Cr in groundwater and to the removal of Cr in water treatment processes. Because of the difference in chemical properties of Cr(III) and Cr(VI), Cr can be mobilized from Cr(III)-containing solids by its oxidation to Cr(VI). Conversely, the reduction of Cr(VI), which is both highly soluble and mobile, can be exploited to remove Cr from potable water. These processes were examined through both laboratory experiments and numerical simulations to assess the geochemical controls on Cr occurrence, speciation, and treatability in groundwater.

EXPERIMENTAL STUDIES

Experimental studies were conducted to investigate the efficiency of Cr(VI) removal from oxygenated waters by ferrous salts, assess the contribution of a colloidal Cr fraction to Cr measured in field samples, test a field method for separation of Cr(III), Cr(VI), and colloidal Cr, and investigate the release of Cr from Cr(III)-containing solids under oxidative and non-oxidative conditions. Based on some of this work, geochemical modeling was performed to investigate both the general aspects of the evolution of the chemical composition of recharge water and a specific example of the effects of sorption on dissolution reactions.

Treatability Study of Redox-Assisted Coagulation

The efficiency of Cr(VI) removal from oxygenated water by ferrous, Fe(II), salts was tested with synthetic solutions and natural Colorado River water (CRW). The contribution of Cr(VI) reduction to its removal was tested by comparing the effectiveness of Fe(II) and Fe(III) salts and by determining the redox speciation of residual dissolved Cr under conditions where Cr removal was incomplete. The effects of pH and solution composition were examined.

- Dissolved Cr(VI) concentrations in both synthetic solutions and natural CRW could be decreased from 50 μg/L to below the detection limit of 0.5 μg/L by addition of Fe(II) salts. Removal efficiency increased with increasing Fe(II) dose. A moderate excess of Fe(II) over the stoichiometric ratio was required for complete Cr removal. Cr removal was less efficient in natural CRW than in simpler synthetic solutions.
- Reduction of Cr(VI) to Cr(III) is a critical step in Cr(VI) removal by Fe(II) salts. Cr(VI) sorption onto Fe(III) oxyhydroxides plays a negligible role in Cr removal, particularly above pH 7.
- Cr(VI) removal by Fe(II) salts was most efficient at pH 7-8. At lower pH values, Cr(VI) removal (over a fixed time interval) was limited by the kinetics of the reaction of Cr(VI) with Fe(II), which decreases with decreasing pH in this range. At higher pH, Cr(VI) removal is limited by the loss of Fe(II) by its competing reaction with dissolved oxygen.

Chromium Speciation

Colloidal Cr

Laboratory studies were conducted to assess the filterability of colloidal Cr formed by supersaturating a solution with amorphous Cr(III) hydroxide at pH 8.2. The contribution of colloidal Cr to Cr measured in field samples was assessed by comparing measured Cr concentrations in filtrates obtained with different size filters.

- Filterability experiments with colloidal suspensions of $Cr(OH)_3(s)$ demonstrated that a significant fraction of Cr passing through 0.2 and 0.02 μm filters was colloidal and could be retained using an ultrafilter with a nominal pore size of 0.003 μm. The Cr concentration measured in the filtrate obtained using the ultrafilter was in reasonable agreement with the predicted concentration of dissolved, inorganic Cr(III) based on equilibrium with the solid.
- In field samples, the consistency of Cr measured in unfiltered, 0.2-μm filtered, and ultrafiltered samples indicated that the contribution of a colloidal fraction was negligible in these samples.

Field methods for Cr speciation

A published method for separation of Cr(III), Cr(VI), and colloidal Cr was tested with solutions containing either Cr(VI) or Cr(III). Performance was acceptable in the former case but not in the latter. Work on this aspect of the project was suspended due to publication by the U.S. Geological Survey of a comparable field method, which was extensively validated against the EPA method for colorimetric determination of Cr(VI).

Dissolution of Cr(III)-containing solids

Weathering studies were conducted with fuchsite, a Cr-containing aluminosilicate mineral, as a model for host silicate phases, amorphous Cr(III) hydroxide as a model for Cr(III) coatings on aquifer minerals, and an aquifer sediment sample obtained from the proposed conjunctive use site. Dissolution was examined under acidic (pH 2) and basic (pH 9) conditions in the presence and absence of the strong oxidant hypochlorous acid/hypochlorite ($HOCl/OCl^-$).

Fuchsite dissolution

Fuchsite dissolution was examined in both batch and flow-through experiments. Rates of Cr, Al, and Si release were quantified under steady-state conditions in flow-through experiments.

- Release of the major constituents Al and Si and the trace constituent Cr were most congruent at pH 9 under oxidizing conditions and least congruent at pH 9 under non-oxidizing conditions. Steady-state release rates (normalized to the elemental composition of the solid) were slower for Cr than for Al and Si under non-oxidizing conditions at both pH 9 and 2.
- Under non-oxidizing conditions, release of Al and Si were not entirely congruent. Preferential release of Al was observed at pH 2 and preferential release of Si at pH 9.

However, rates of release of these major constituents were generally consistent with values reported in the literature for muscovite dissolution.

- Under oxidizing conditions, Si and Al were released somewhat more slowly than under non-oxidizing conditions. It is possible that the oxidation of Fe(II) in the silicate matrix contributed to this effect but this hypothesis remains to be tested.

$Cr(OH)_3(s)$ dissolution

Dissolution of $Cr(OH)_3(s)$ was examined in batch experiments. Sorption of Cr(VI) onto $Cr(OH)_3(s)$ was examined as a function of pH and as a function of dissolved Cr(VI) concentration at pH 3.

- At pH 9, dissolved Cr concentrations increased linearly with time corresponding to dissolution of 40% of the solid in 200 min. Release of Cr was negligible under non-oxidizing conditions at this pH.
- At pH 2, some dissolution of the solid was observed under oxidizing conditions (though much less than at pH 9). The rate of the reaction at pH 2 decreased with time, however, and the dissolved Cr concentration appeared to reach a plateau. Under non-oxidizing conditions, Cr release was more nearly linear but also slower than under oxidizing conditions.
- Sorption of Cr(VI) onto $Cr(OH)_3(s)$ is strongly pH dependent with negligible sorption at pH 9 and increasing sorption with decreasing pH. The sorption isotherm obtained at pH 3 suggests that, under the conditions of the dissolution experiment at pH 2 under oxidizing conditions, the $Cr(OH)_3(s)$ surface would rapidly become saturated with Cr(VI). This may inhibit the further oxidative dissolution of the solid.

Dissolution of aquifer sediment. Release of Al, Si, and Cr from aquifer sediments was examined in batch experiments only at pH 9 but under both oxidizing and non-oxidizing conditions. The Cr content of the aquifer sediment was at the detection limit for the XRF analysis (0.01%). Dissolution rates could not be quantified under non-oxidizing conditions.

- Under oxidizing conditions, release of Cr from the solid was observed despite its low Cr content. Cr concentrations increased linearly with time but reached only 4 μg/L in 14 d.

Geochemical Modeling

Geochemical modeling was performed to examine the evolution of water composition with time in two cases. In the first case, dissolution of a primary solid was considered alone or with precipitation-dissolution of a secondary solid and/or sorption-desorption of the dissolved constituent. In the second case, oxidative dissolution of $Cr(OH)_3(s)$ was considered with subsequent adsorption of the product Cr(VI); sorbed Cr(VI) was also allowed to inhibit the further oxidative dissolution of the solid.

- Dissolution of the primary solid results in the linear increase in the dissolved concentration of its constituent elements (here called X). Precipitation-dissolution of a secondary solid (also containing X) results in the dissolved concentration of X reaching a steady-state value (even as the dissolution of the primary solid continues). Sorption-desorption of X (under conditions such that the concentration of surface sites are limited) slows the accumulation of

X in the dissolved phase but does not change the eventual steady-state concentration of X (which is determined by pseudo-equilibrium with the secondary solid).

- In the oxidative dissolution of $Cr(OH)_3(s)$, rapid saturation of surface sites is predicted (i.e., for pH conditions favoring Cr(VI) adsorption). Inclusion of an inhibitory effect of sorbed Cr(VI) provides a reasonable model for the observed release of Cr from $Cr(OH)_3(s)$ under oxidizing conditions at pH 2. Sorption alone (i.e., without inhibition) has little effect on the accumulation of Cr in solution once the surface is saturated.

NUMERICAL SIMULATIONS

Soil and groundwater flow and transport simulations using the numerical models FEMWATER and HGBC123D were used to evaluate the potential field-scale effects of natural chromium sources on ASR operations. Two sets of model simulations were undertaken. The first and more extensive simulations employed a single computational domain to model the unsaturated and saturated zones simultaneously. A second set of simulations was undertaken to test the prospect of decoupling the unsaturated and saturated zone simulations in order to provide a more detailed account of the unsaturated zone while alleviating computational burden. In all simulations, ASR water quality was estimated for local (recovery well) behavior, and global behavior as summarized by spatial moments estimates of the simulated spatial chromium concentration distributions.

Single Domain Simulations

In the single-domain simulations, three scenarios were tested: (I) Preexisting background Cr concentrations in the aquifer (before ASR operations begin) but no dissolution/weathering source, (II) Preexisting Cr concentrations and zero-order dissolution/weathering sources uniformly distributed in the aquifer, and (III) Zero-order dissolution/weathering in the unsaturated zone only. For Scenario I, key parameters were identified and selected for testing the sensitivity ASR efficiency in terms of recovered water quality. These parameters included dispersivity, recharge rate and recharge configuration. Scenario II focused on the impact of the weathering rate on recovered water quality, invoking the range of zero-order rate constants gleaned from the literature on silicate weathering (Chapter 1). Scenario III again focused on the weathering rate constants, but from the perspective of the potential impact of unsaturated zone weathering products on recovered water quality.

- The advantage of the single-domain approach is that it captures the true dynamics of the unsaturated and saturated flow processes. More specifically, the single-domain approach captures the mounding dynamics which occur when the recharging water reaches the water table. The mounding in turn dominates local vertical and horizontal flow patterns in the aquifer.
- Parameter sensitivity analyses indicated that the greatest changes in the simulated amount of clean water recovered (i.e., ASR efficiency) resulted from changes in recharge rate and recharge configuration. This result is a function of the mounding effect on mixing between the clean (recharged) and native groundwater. The horizontal movement and trajectory of the clean water plume was most affected by recharge rate and duration of the recharge period. Concentration time series at observation wells downgradient from the recharge zone were most sensitive to the value of dispersivity.

- For the domain and conditions simulated, the impact of dissolution rate on normalized local concentrations downgradient are affected by the value of incoming concentration. Incoming concentration had a significant effect at the low reaction rates, but the effect lessened with increasing reaction rate. There was an upper bound for the normalized concentration when dissolution into the recharged water took place against a background that had reached steady state. Cr weathering/dissolution rates of 1e-12 mol/kg/d and lower were sufficiently slow to produce output that closely resembled that for the non-reactive transport base. This means that the dissolution kinetics could be negligible over the time scale associated with these cases. For reaction rates of 1e-9 mol/kg/d and greater, local equilibrium was valid throughout the simulated domain. Intermediate rates (1e-12 mol/kg/d < k < 1e-9 mol/kg/d) produced significant effects on the recovered water quality for the conditions simulated. The results from the literature suggest that silicate weathering rates are extremely low, ranging from 10^{-17} to 10^{-11} mol/m^{2}s. Given that the stoichiometric content of chromium in most silicate minerals is expected to relatively low (< 1%), congruent chromium dissolution rates may be two or more orders of magnitude lower than this range. These literature values suggest that Cr weathering would not impact ASR water significantly for the geometry and timescales studied (roughly 10-15 year storage periods). However, mixing with the native water will always impact the ASR water quality.
- The application of spatial moment analysis was demonstrated in this work as a robust method for quantifying clean plume behavior in the context of an ASR setting. Spatial moments provide an integrated assessment of simulated ASR efficiency or clean-water recovery (zeroth moment), clean-water plume centroid location (first moments with respect to different axes), and clean-water plume travel time (first moment derivative with time).
- The disadvantage of the single-domain approach (and motivation for the dual-domain approach) is that accuracy of the simulation is highly suspect in the absence of a high-resolution spatial and temporal discretization in the unsaturated portion of the domain. The problem with this is that the overall scale of the combined saturated and unsaturated grids leads to excessive computation burden. The greater computational requirements are attributed to the highly nonlinear form the governing equations take under unsaturated conditions. These nonlinearities are handled by increasing both the spatial and temporal resolutions, thus increasing the computational time required for a simulation. In the simulations used here, the "fine" resolution simulations were actually coarse in an absolute sense to allow simulation times ranging from 30 min to a few hours. However, trial simulations focusing on detail in the unsaturated zone drove these simulation times up to tens of hours or even days on PC Pentium 4 processors operating at 2.5 GHz.

Dual Domain Simulations

The dual-domain simulations required numerical innovations and were undertaken using the same parameters as in the base simulations for the single-domain approach. The main innovation was associated with developing a method for feeding the flux from an unsaturated (vadose) zone simulation at high-resolution to a large-scale, low-resolution grid in the saturated (groundwater) domain. The resulting numerical scheme for matching the grids at the unsaturated-saturated interface is referred to as the recharge boundary construction algorithm.

- The dual-domain simulations based on the boundary recharge construction algorithm were successful if the value of the key numerical error criterion (maximum % difference) selected was sufficiently small. A tradeoff was identified between this criterion and the computational times required for the dual-domain simulations. For the domains and conditions of this study, computation times ranged from greater than 10 to less than 3 hours with decreasing levels of stringency. Thus, the method is viable for large scale systems if computational times on the order of several hours are acceptable. Parallel processing algorithms and faster computers can be used to reduce computation times.

CHAPTER 7
RECOMMENDATIONS TO UTILITIES

Groundwater storage offers substantial benefits to utilities that seek additional flexibility in managing their water supply. However the use of this water management strategy may be compromised if the quality of recovered water is degraded relative to that of the water used for recharge. In some cases, groundwater storage will only be feasible if the recovered water can be distributed to consumers with only minimal treatment (e.g., disinfection). Even when recovered water is delivered to a treatment facility prior to distribution, some concerns may arise if water quality is degraded during storage in the subsurface.

This project addressed several issues related to the occurrence and treatability of chromium in groundwater through laboratory and field experiments and numerical simulations. On the basis of the experimental studies, the following recommendations can be made:

- Chromium(VI) concentrations in ambient groundwater or recovered water can be decreased to sub-ppb levels by addition of Fe(II) salts even in the presence of dissolved oxygen. The mechanistic basis for this redox-assisted coagulation process is the reaction of Cr(VI) with Fe(II) to produce Cr(III) and Fe(III). Hydrolysis and precipitation of the products results in incorporation of Cr(III) in a mixed Cr(III)-Fe(III) hydroxide solid, which can be removed by filtration. Treatment efficiency can be improved by pH optimization and by increasing the Fe(II) dose. These tests were run on waters spiked with Cr(VI) and would need to be validated for source waters containing elevated Cr concentrations. Another recent study has reported residual Cr concentrations of 2-3 μg/L after treatment with Fe(II) (Brandhuber et al., 2004). Although our tests were not run in the absence of dissolved oxygen, previous studies demonstrate that the process would also be effective under those conditions.
- The presence of colloidal Cr(III) could be detected in laboratory studies by comparing the total Cr concentrations measured in ultrafiltered samples with those measured in filtrates passed through 0.2 μm filters. A similar test run with samples collected in the field demonstrated that colloidal Cr was not present in these field samples. Assessment of the possible contribution of colloidal Cr would be particularly important for water samples in which the total Cr concentrations in conventionally filtered samples exceeded the measured Cr(VI) concentrations.
- The rate of Cr release from aquifer minerals will be influenced by whether the oxidation of Cr(III) to Cr(VI) occurs concurrent with or subsequent to mineral dissolution. The extent of this effect will depend on the type of mineral hosting Cr in the soil or aquifer sediment. Cr release from silicate minerals is very slow and is congruent with the release of the major elements Al and Si only under oxidizing conditions. Under non-oxidizing conditions, Cr release is even slower than the dissolution of the mineral framework. If Cr is present as hydroxide coatings on aquifer minerals, however, its release will be faster than if is hosted in aluminosilicate minerals. Release of Cr from Cr(III)hydroxide solids is strongly enhanced under oxidizing conditions at basic pH. At acidic pH, however, the product Cr(VI) adsorbs onto the residual solid, which appears to inhibit further dissolution.
- Simple geochemical modeling can be used to illustrate the effects of dissolution-precipitation and sorption-desorption on the accumulation of dissolved constituents over time under batch reactor conditions. In particular, sorption can potentially influence the rate of accumulation of a dissolved constituent, either through its loss from solution to sorption sites or the inhibition of the reaction of the underlying solid. However, if the concentration of sorption

sites is limited, the steady-state concentration of dissolved constituents will not ultimately be limited by sorption. The case of zero-order dissolution of primary minerals was further examined in the context of detailed transport modeling.

The following recommendations can be made on the basis of the numerical studies:

- Any proposed ASR project should be accompanied by a modeling investigation aimed at predicting flow and transport effects under anticipated operating conditions. The general procedures for model construction, discretization and parameter sensitivity analysis are described in this work, but other ASR projects will undoubtedly be subject to different scales and conditions and may have different operational horizons.
- Single-domain ASR simulations are a viable modeling alternative if resolution of results in the unsaturated zone is not important. This approach is preferable, especially if the simulation model is to be coupled to a water resources management algorithm requiring repetitive simulations. Cases clearly exist where better unsaturated zone resolution is needed. For example, proximate groundwater-surface water interactions, such as at streams or springs, may be impacted by the recharge and extraction rates, and by input/output timing of the ASR operation.
- If resolution in the unsaturated zone is a requirement, or if the model is to also serve as a water resources management tool (placing the model within the framework of an optimization algorithm), then the additional computational costs associated with a high resolution single-domain model would become unacceptable. In this case it is better to explore the use of the dual-domain approach described in this study.
- The potential impact of the natural chromium release rates likely to govern in the case of silicate weathering processes were shown here to be minimal for the conditions and timescales simulated here. This means that release rates identified in the literature were sufficiently slow relative to the clean-water residence time in this system. This result means that conservative transport simulation using FEMWATER would suffice and the more complex HBGC123D model could be avoided. It is important to note that this result is linked to the scale of the ASR system and the ASR operational parameters governing the clean-water residence time. An ASR project with a greater residence time, faster weathering rates, or some combination of the two, may demonstrate significant chromium introduction into clean-water plumes during the storage cycle of an ASR operation.

APPENDIX A
SHORT-TERM LEACHING EXPERIMENTS

Leaching studies were performed with fuchsite and $Cr(OH)_3$ samples in acidic conditions with three different acids or basic conditions with a strong oxidant (NaOCl). Standard procedures for chromium extraction from soil samples (Bartlett, 1991) were followed with only slight modifications. Chromium in both fuchsite and $Cr(OH)_3$ exists as trivalent Cr, Cr(III). The amount of Cr extracted with acids or a strong oxidant under basic conditions is generally considered to represent the total extractable or total oxidizable Cr, respectively. Based on the results presented below, however, short-term leaching, even with high concentrations of acids or a strong oxidant, was found to extract only a small fraction of Cr, Al, and Si from fuchsite samples. These methods are clearly insufficient to assess total elemental composition as has been recommended in the literature (Bartlett, 1991).

LEACHING PROCEDURE

Three different acidic media (one mineral acid and two organic acids) were used for acid leaching experiments: 0.1 M HNO_3, 0.01 M NaH_2-citrate (pH 3.70) and 0.01 M oxalic acid adjusted to pH 3.73 with 1.0 M NaOH. 200 mg of powdered fuchsite samples (> 125 μm) or 10 mg of $Cr(OH)_{3(s)}$ was transferred to an extraction vessel (15 mL polyethylene bottles) and 10 mL of an acidic medium was added. The mixture was shaken in an end-over-end shaker operating at 40 rpm for 18 h at 25.0 °C. The extracts were separated by filtration using 0.22 μm filters and were analyzed for the concentrations of Cr, Al and Si by ICP-MS after acidification to a final concentration of 1% with HNO_3.

Total oxidizable Cr(III) was also determined with a strong oxidizing medium. 500 mg of powdered fuchsite samples (>125 μm) or 25 mg of $Cr(OH)_{3(s)}$ was transferred to an extraction tube (50 mL polyethylene bottles) and 25 mL of 0.97 M NaOCl adjusted to pH 9.5 with 5.0 M HNO_3 was added. The mixtures were heated at 80 - 90 °C for 1 hr in a water bath on a hot plate and were cooled at room temperature. The extracts were obtained and analyzed as described above.

RESULTS OF LEACHING UNDER ACIDIC CONDITIONS

Only a small fraction (ca. 0.04% excluding sample F5) of the total Cr was extracted from the fuchsite samples (>125 μm fraction) by the acid media. A smaller fraction of total Cr was extracted from sample F5, reflecting the higher total Cr concentration but an extracted Cr concentration comparable to the other samples (Table A.1). Similar Cr concentrations were obtained by extraction with all three acids and with all samples (Fig. A.1). The extracted Cr were 0.030 ± 0.014 μmol/g with HNO_3, 0.038 ± 0.021 μmol/g with citric acid and 0.026 ± 0.024 μmol/g with oxalic acid (Table A.2). Extractable Al concentrations were almost identical in the three acidic media and were 2.8 ± 0.7, 3.0 ± 0.7 and 3.2 ± 0.5 μmol/g with HNO_3, citric acid and oxalic acid, respectively. For Si, the extractable concentrations with HNO_3 and citric acid were similar but consistently lower with oxalic acid than with the other two acidic media. In addition, the concentrations of extractable Si were consistently lower than those of Al despite the higher total concentrations in fuchsite, indicating the preferential release of Al over Si from the solid (Fig. A.1).

The fraction of the extracted Cr with the acids was 0.04 ± 0.03% of the total Cr in the solid (Fig. A.2). Likewise, only a small fraction of Al (0.05%) and Si (0.02%) of their total contents were extracted from the fuchsite samples by the acid media. The concentrations of extractable Cr, Al, and Si were generally four orders of magnitude lower than their total concentrations of the fuchsite samples.

The concentrations of the extractable Al were about two orders of magnitude higher than those of Cr, and the ratio of the concentrations of extractable Cr to Al were 0.011 ± 0.003, 0.012 ± 0.005 and 0.007 ± 0.006 with HNO_3, citric acid and oxalic acid, respectively (Fig. A.3). By comparing the ratios of the concentrations of extractable Si to Al, those with HNO_3 and citric acid were identical but consistently lower with oxalic acid (0.29 ± 0.06) than those with the other two acids. The ratio of Cr/Al in the acid extracts (0.011 mol Cr/mol Al) was slightly lower than that in the original solid (0.018 mol Cr/mol Al). Similarly, the ratio of Si/Al in the acid extracts (0.40 mol Si/mol Al) was also lower than that in the original solid (1.24 mol Si/mol Al). These results indicate that Al is preferentially released from fuchsite over Cr and Si under acidic conditions and are consistent with the results from fuchsite dissolution under non-oxidizing and acidic conditions (cf. Fig. 4.2b).

By contrast, the extractable Cr concentrations in $Cr(OH)_{3(s)}$ were significantly higher than those in the fuchsite samples and were 1,600 μmol/g with HNO_3, 280 μmol/g with citric acid and 1,200 μmol/g with oxalic acid, which accounted for 16%, 3% and 12%, respectively, of the total Cr concentrations of $Cr(OH)_{3(s)}$ (Fig. A.4).

RESULTS OF LEACHING IN OXIDIZING CONDITIONS

Chromium in fuchsite was not extracted under oxidizing conditions and its extractable concentration was below detection limit in every case (Table A.1). However, $Cr(OH)_{3(s)}$ was completely dissolved under these oxidizing conditions (Fig. A.4). These results indicate that Cr in silicate minerals is very resistant to dissolution compared to that in oxides or hydroxides. Consistent with this observation, Cr in the residual fraction of chromite ore processing residues has been shown to be relatively unreactive (as determined by extraction with 40% (w/v) H_2O_2 and concentrated HNO_3) compared to the other reactions (Sreeram and Ramasami, 2001).

Table A.1
Total elemental compositions of fuchsite and $Cr(OH)_3(s)$

Sample ID	Cr mmol/g	Al mmol/g	Si mmol/g
F1	0.093	6.3	7.7
F2	0.063	6.6	7.4
F3	0.044	5.6	7.9
F4	0.064	6.2	7.8
F5	1.4	5.8	4.9
F6	0.13	6.1	7.6
$Cr(OH)_{3(s)}$	9.7	—	—

Table A.2
Concentrations of elements released in leaching experiments

Extractant	sample*	Cr		Al		Si	
		μmol/g	%	μmol/g	%	μmol/g	%
HNO_3	F1	0.034	0.04	2.4	0.04	1.3	0.02
(0.1 M)	F2	0.012	0.02	1.6	0.02	1.4	0.02
	F3	0.031	0.07	3.1	0.06	1.6	0.02
	F4	0.018	0.03	2.8	0.04	1.4	0.02
	F5	0.044	0.003	3.0	0.05	1.4	0.03
	F6	0.046	0.04	3.8	0.06	1.8	0.02
	$Cr(OH)_{3(s)}$	1,600	16	—		—	
Citric acid	F1	0.037	0.04	2.6	0.04	1.3	0.02
(0.01 M,	F2	0.012	0.02	2.2	0.03	1.1	0.02
pH 3.70)	F3	0.047	0.11	3.1	0.05	1.5	0.02
	F4	0.019	0.03	3.1	0.05	1.3	0.02
	F5	0.038	0.003	2.6	0.04	1.4	0.03
	F6	0.072	0.06	4.1	0.08	2.0	0.03
	$Cr(OH)_{3(s)}$	280	2.9	—		—	
Oxalic acid	F1	0.021	0.02	2.6	0.04	0.77	0.01
(0.01 M,	F2	< 0.005	< 0.008	2.7	0.04	0.75	0.01
pH 3.73)	F3	0.033	0.08	3.2	0.06	0.95	0.01
	F4	0.005	0.008	3.3	0.05	0.67	0.008
	F5	0.030	0.002	3.0	0.05	0.86	0.02
	F6	0.066	0.05	4.0	0.06	1.5	0.02
	$Cr(OH)_{3(s)}$	1,200	12	—		—	
NaOCl	F1	< 0.005		< 0.05		< 0.05	
(0.97 M,	F2	< 0.005		< 0.05		< 0.05	
pH 9.5)	F3	< 0.005		< 0.05		< 0.05	
	F4	< 0.005		< 0.05		< 0.05	
	F5	< 0.005		< 0.05		< 0.05	
	F6	< 0.005		< 0.05		< 0.05	
	$Cr(OH)_{3(s)}$	11,000	113	—		—	

*F1-F6: fuchsite

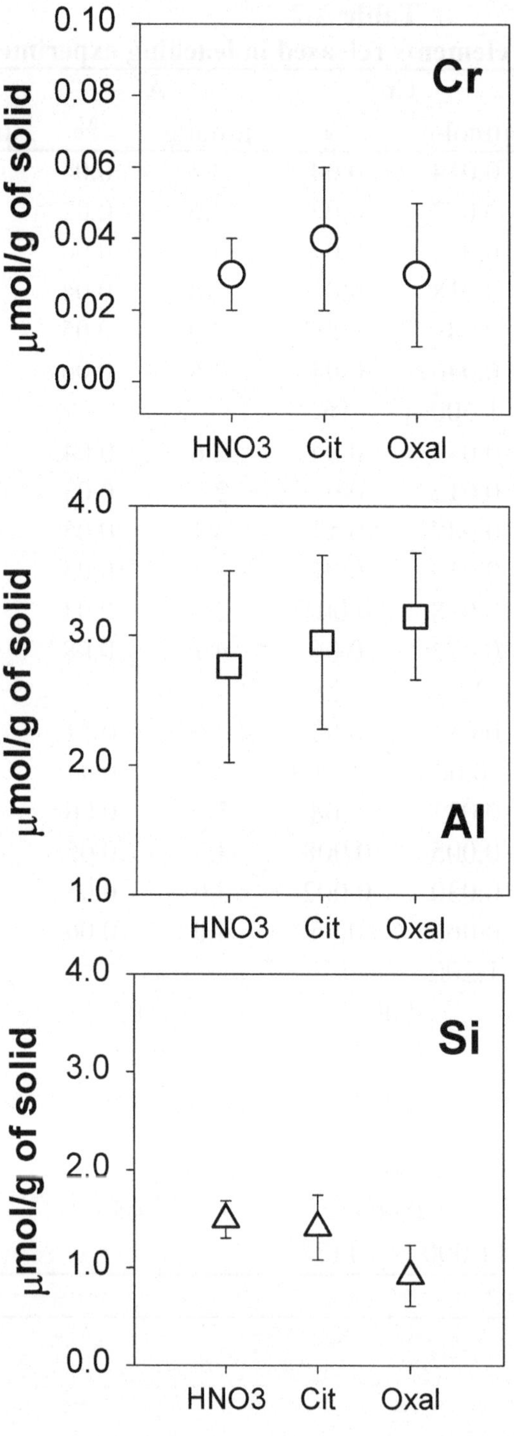

Figure A.1 Amounts of metals leached from fuchsite with three different acids. Error bars indicate the standard deviation of the six samples.

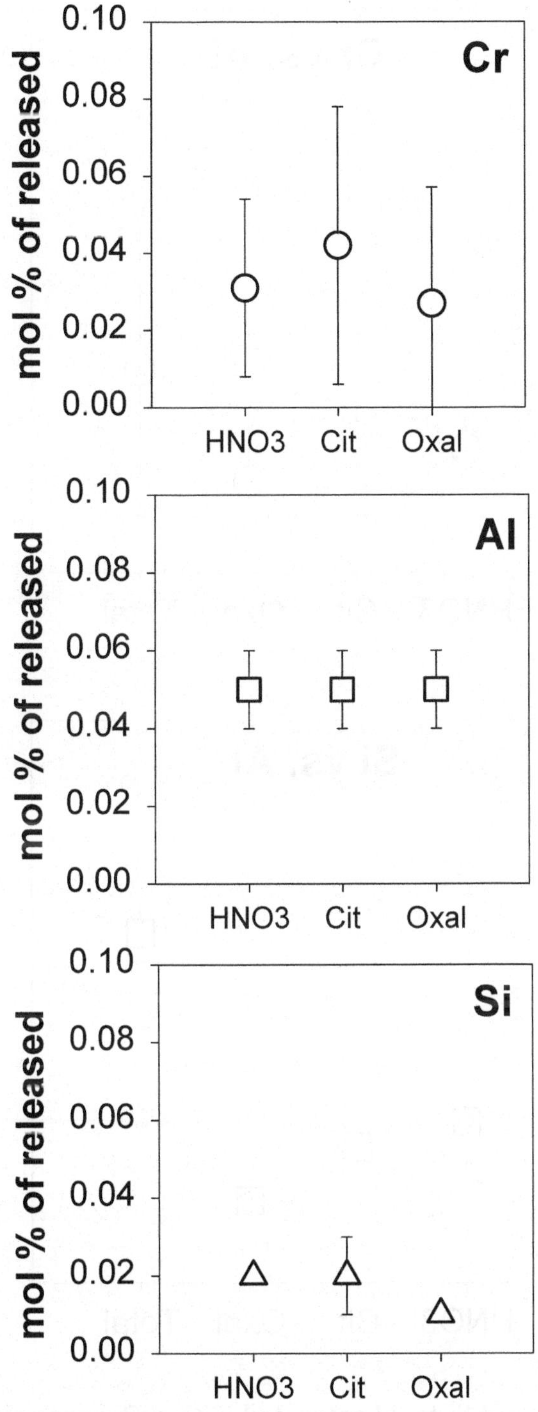

Figure A.2 Molar fraction of Cr, Al, and Si extracted from fuchsite samples with acids. Error bars as in Figure A.1.

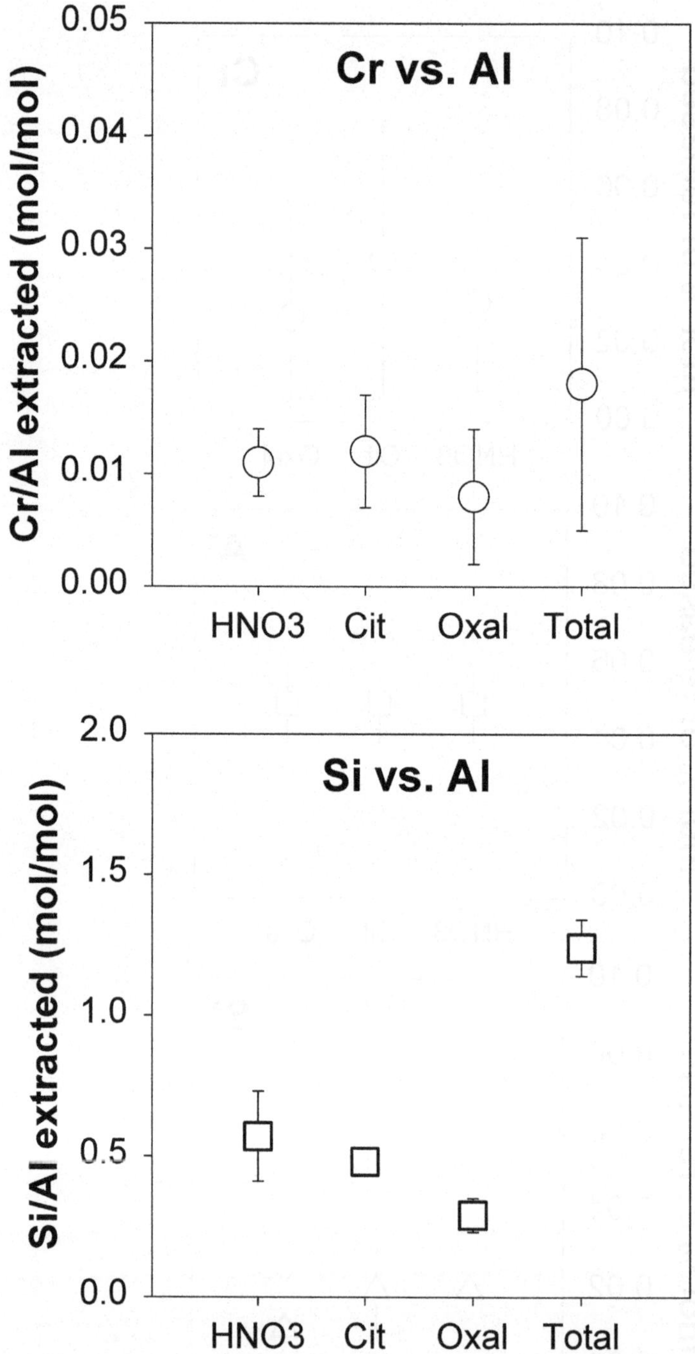

Figure A.3 Molar ratio of Cr and Si to Al extracted with different acids and those in the original solid. Error bars as in Figure A.1.

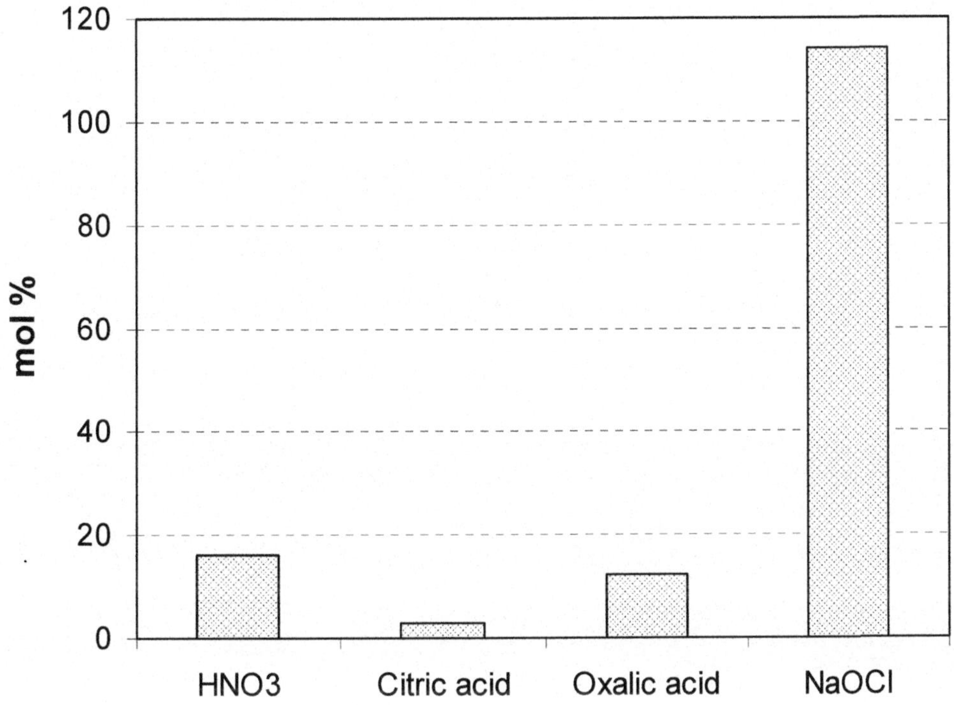

Figure A.4 Fraction of Cr extracted from $Cr(OH)_3$ with different acids and a strong oxidant.

REFERENCES

Aktor, H., 1994. Continuous High-Rate Removal of Chromate in a Fluidized-Bed without Sludge Generation. *Water Science and Technology*, 30(7): 31-40.

Anderson, R.A., 2000. Chromium in the Prevention and Control of Diabetes. *Diabetes & Metabolism*, 26(1): 22-27.

Aoki, T. and M. Munemori, 1982. Recovery of Chromium(VI) from Wastewaters with Iron(III) Hydroxide - I. Adsorption Mechanism of Chromium(VI) on Iron(III) Hydroxide. *Water Res.*, 16: 793-796.

Arar, E.J. and J.D. Pfaff, 1991. Determination of Dissolved Hexavalent Chromium in Industrial Waste-Water Effluents by Ion Chromatography and Postcolumn Derivatization with Diphenylcarbazide. *Journal of Chromatography*, 546(1-2): 335-340.

Arthur, J.D., A.A. Dabous and J.B. Cowart, 2002. Mobilization of Arsenic and Other Trace Elements During Aquifer Storage and Recovery, Southwest Florida. In: G.R. Aiken and E.L. Kuniansky (Editors), U.S. Geological Survey Artifical Recharge Workshop Proceedings, Sacramento, Ca, April 2-4, 2002. U.S. Geological Survery Open-File Report 02-89.

ATSDR, 2003. Health Consultation: Total and Hexavalent Chromium, Lead,Mercury, Nitrate, and Thallum Levels in Water from Private Wells near the Pacific Gas and Electric Facility in Hinkley, CA. Agency for Toxic Substances and Disease Registry, http://www.atsdr.cdc.gov/HAC/PHA/pgande/pgf_toc.html.

Ball, J.W. and R.B. McCleskey, 2003. A New Cation-Exchange Method for Accurate Field Speciation of Hexavalent Chromium. WRI 03-4018, U.S. Geological Survey Water Resources Investigations.

Ball, J.W. and D.K. Nordstrom, 1998. Critical Evaluation and Selection of Standard State Thermodynamic Properties for Chromium Metal and Its Aqueous Ions, Hydroloysis Species, Oxides, and Hydroxides. *J. Chem. Eng. Data*, 43: 895-918.

Bartlett, R.J., 1991. Chromium Cycling in Soils and Water - Links, Gaps, and Methods. *Environmental Health Perspectives*, 92: 17-24.

Beaubien, S., J. Nriagu, D. Blowes and G. Lawson, 1994. Chromium Speciation and Distribution in the Great-Lakes. *Environmental Science & Technology*, 28(4): 730-736.

Benoit, G. and T.F. Rozan, 1999. The Influence of Size Distribution on the Particle Concentration Effect and Trace Metal Partitioning in Rivers. *Geochimica Et Cosmochimica Acta*, 63(1): 113-127.

Berner, R.A. and J. Holdren, G. R., 1979. Mechanism of Feldspar Weathering - Ii. Observations of Feldspars from Soils. *Geochim. Cosmochim. Acta*, 43: 1173-1186.

Blankstein, A. and J. Guccione, 2000. Identifying Chromium 6 Sites Could Take Months, Los Angeles Times, Los Angeles, October 13, p. B3.

Brady, P.V. and J.V. Walther, 1989. Controls on Silicate Dissolution Rates in Neutral and Basic Ph Solutions at 25 °C. *Geochim. Cosmochim. Acta*, 53: 2823-2830.

Brandhuber, P., N. Blute, C. Seidel, M.M. Frey, J. Yoon, G. Amy and D. Froelich, 2004. Alternative Treatment Technologies for the Control of Hexavalent Chromium to Very Low Levels, AWWA Inorganics Workshop, Reno, NV.

Brigatti, M.F., E. Galli, L. Medici, L. Poppi, G. Cibin, A. Marcelli and A. Mottana, 2001. Chromium-Containing Muscovite: Crystal Chemistry and XANES Spectroscopy. *Eur. J. Mineral.*, 13: 377-389.

Buerge, I.J. and S.J. Hug, 1997. Kinetics and Ph Dependence of Chromium(VI) Reduction by Iron(II). *Environmental Science & Technology*, 31(5): 1426-1432.

Buerge, I.J. and S.J. Hug, 1999. Influence of Mineral Surfaces on Chromium(VI) Reduction by Iron(II). *Environmental Science & Technology*, 33(23): 4285-4291.

Byrdy, F.A., L.K. Olson, N.P. Vela and J.A. Caruso, 1995. Chromium Speciation by Anion-Exchange High-Performance Liquid- Chromatography with Both Inductively-Coupled Plasma-Atomic Emission Spectroscopic and Inductively-Coupled Plasma-Mass Spectrometric Detection. *Journal of Chromatography A*, 712(2): 311-320.

Cardelli, A., G. Cibin, S. Benfatto, S. Della Longa, M.F. Brigatti and A. Marcelli, 2003. A Crystal-Chemical Investigation of Cr Substitution in Muscovite by XANES Spectroscopy. *Phys. Chem. Minerals*, 30: 54-58.

Carroll, S.A. and J.V. Walther, 1990. Kaolinite Dissolution at 25°, 60° and 80°C. *Amer. J. Sci.*, 290: 797-810.

Casey, W.H., J.F. Banfield, H.R. Westrich and L. McLaughlin, 1993. What Do Dissolution Experiments Tell Us About Natural Weathering. *Chem. Geol.*, 105: 1-15.

Casey, W.H. and H.R. Westrich, 1992. Control of Dissolution Rates of Orthosilicate Minerals by Divalent Metal-Oxygen Bonds. *Nature*, 355: 157-159.

Collins, C.H., S.H. Pezzin, J.F.L. Rivera, P.S. Bonato, C.C. Windmoller, C. Archundia and K.E. Collins, 1997. Liquid Chromatographic Separation of Aqueous Species of Cr(VI) and Cr(III). *Journal of Chromatography A*, 789(1-2): 469-478.

Cox, A.G., I.G. Cook and C.W. McLeod, 1985. Rapid Sequential Determination of Chromium(III) - Chromium(VI) by Flow-Injection Analysis - Inductively Coupled Plasma Atomic- Emission Spectrometry. *Analyst*, 110(4): 331-333.

Cox, A.G. and C.W. McLeod, 1986. Preconcentration and Determination of Trace Chromium(III) by Flow-Injection Inductively-Coupled Plasma/Atomic Emission-Spectrometry. *Analytica Chimica Acta*, 179: 487-490.

Cox, A.G. and C.W. McLeod, 1992. Field Sampling Technique for Speciation of Inorganic Chromium in Rivers. *Mikrochimica Acta*, 109(1-4): 161-164.

Cutter, G.A., 1992. Kinetic Controls on Metalloid Speciation in Seawater. *Marine Chemistry*, 40: 65-80.

Davis, J.A., D.B. Kent, J.A. Coston, K.M. Hess and J.L. Joye, 2000. Multispecies Reactive Tracer Test in an Aquifer with Spatially Variable Chemical Conditions. *Water Resources Research*, 36(1): 119-134.

Davis, J.A. and J.O. Leckie, 1980. Surface Ionization and Complexation at the Oxide/Water Interface. 3. Adsorption of Anions. *J. Colloid Int. Sci.*, 74: 32-43.

DHS, 2003. Chromium-6 in Drinking Water. CA Dept. of Health Services, http://www.dhs.cahwnet.gov/ps/ddwem/chemicals/chromium6/Cr+6backgroundinfo.htm #Health%20Concerns%20about%20Cr+6.

Domenico, P.A. and F.W. Schwartz, 1998. *Physical and Chemical Hydrogeology*. John Wiley & Sons, New York, 506 pp.

Drever, J.I., 1997. *The Geochemistry of Natural Waters*. Prentice-Hall, Upper Saddle River, NJ, 436 pp.

Dzombak, D.A. and F.M. Morel, 1990. *Surface Complexation Modeling: Hydrous Ferric Oxide*. Wiley-Interscience, New York, 393 pp.

Eary, L.E. and D. Rai, 1987. Kinetics of Chromium(III) Oxidation to Chromium(VI) by Reaction with Manganese-Dioxide. *Environmental Science & Technology*, 21(12): 1187-1193.

Eary, L.E. and D. Rai, 1988. Chromate Removal from Aqueous Wastes by Reduction with Ferrous Ion. *Environmental Science & Technology*, 22(8): 972-977.

Eggleston, C.M., J. Hochella, M. F. and G.A. Parks, 1989. Sample Preparation and Aging Effects on the Dissolution Rate and Surface Composition of Diopside. *Geochim. Cosmochim. Acta*, 53: 797-804.

Elison, D., 2003. Status of Aquifer Storage and Recovery Issues and Research in the Southwest Florida Water Management District, 2003 Annual Water Resources Conference. American Water Resources Association, San Diego, CA.

Elovitz, M.S. and W. Fish, 1994. Redox Interactions of Cr(VI) and Substituted Phenols - Kinetic Investigation. *Environmental Science & Technology*, 28(12): 2161-2169.

El-Shoubary, Y., N. Speizer, S. Seth and H. Savoia, 1998. A Pilot Plant to Treat Chromium-Contaminated Groundwater. *Environmental Progress*, 17(3): 209-213.

Fantoni, D., G. Brozzo, M. Canepa, F. Cipolli, L. Marini, G. Ottonello and M.V. Zuccolini, 2002. Natural Hexavalent Chromium in Groundwaters Interacting with Ophiolitic Rocks. *Environmental Geology*, 42(8): 871-882.

Farias, S.S., V.A. Casa, C. Vazquez, L. Ferpozzi, G.N. Pucci and I.M. Cohen, 2003. Natural Contamination with Arsenic and Other Trace Elements in Ground Waters of Argentine Pampean Plain. *Science of the Total Environment*, 309(1-3): 187-199.

Fendorf, S., M.J. Eick, P. Grossl and D.L. Sparks, 1997. Arsenate and Chromate Retention Mechanisms on Goethite .1. Surface Structure. *Environmental Science & Technology*, 31(2): 315-320.

Fendorf, S.E., M. Fendorf, D.L. Sparks and R. Gronsky, 1992b. Inhibitory Mechanisms of Cr(III) Oxidation by Delta-MnO_2. *Journal of Colloid and Interface Science*, 153(1): 37-54.

Fendorf, S.E. and G.C. Li, 1996. Kinetics of Chromate Reduction by Ferrous Iron. *Environmental Science & Technology*, 30(5): 1614-1617.

Fendorf, S.E. and R.J. Zasoski, 1992a. Chromium(III) Oxidation by Delta-MnO_2 .1. Characterization. *Environmental Science & Technology*, 26(1): 79-85.

Flegal, R., J. Last, E.E. McConnell, M. Schenker and H. Witschi, 2001. Scientific Review of Toxicological and Human Health Issues Related to the Development of a Public Health Goal for Chromium(VI), http://www.dhs.cahwnet.gov/ps/ddwem/chemicals/chromium6/reviewpanelreport.pdf.

FR, 1995. Guidelines Establishing Test Procedures for the Analysis of Pollutants: New Methods; Proposed Rule. Vol. 60, No. 201, Federal Register, Washington, D.C.

Freeze, R.A. and J.A. Cherry, 1979. *Groundwater*. Prentice-Hall, NJ, 604 pp.

Freyberg, D.L., 1986. A Natural Gradient Experiment on Solute Transport in a Sand Aquifer 2. Spatial Moments and the Advection and Dispersion of Nonreactive Tracers. *Water Resources Research*, 22: 2031-2046.

Gaus, I., P. Shand, I.N. Gale, A.T. Williams and J.C. Eastwood, 2002. Geochemical Modelling of Fluoride Concentration Changes During Aquifer Storage and Recovery (ASR) in the Chalk Aquifer in Wessex, England. *Quarterly Journal of Engineering Geology and Hydrogeology*, 35: 203-208.

Gelhar, L.W., C. Welty and K.R. Reyfeldt, 1992. A Critical Review of Field-Scale Dispersion in Aquifers. *Water Resources Research*, 28: 1955-1974.

Giammar, D.E. and J.G. Hering, 2002. Equilibrium and Kinetic Aspects of Soddyite Dissolution and Secondary Phase Precipitation in Aqueous Suspension. *Geochim. Comoschim. Acta*, 66: 3235-3245.

Grenthe, I., W. Stumm, M. Laaksuharju, A.C. Nilsson and P. Wikberg, 1992. Redox Potentials and Redox Reactions in Deep Groundwater Systems. *Chemical Geology*, 98: 131-150.

Guccione, J., 2000. Burbank Pollution Dogs Lockheed Years Later, Los Angeles Times, Los Angeles, October 5, p. A1.

Gwo, J.P., H. Frenzel, E. D'Azevedo and F.M. Hoffman, 1999. Hydrobiogeochem123d, V. 1.1, Oak Ridge National Lab.

Henderson, T., 1994. Geochemical Reduction of Hexavalent Chromium in the Trinity Sand Aquifer. *Ground Water*, 32: 477-486.

Hoch, A.R., M.M. Reddy and J.I. Drever, 1996. The Effect of Iron Content and Dissolved O_2 on Dissolution Rates of Clinopyroxene at pH 5.8 and 25°C: Preliminary Results. *Chem. Geol.*, 132: 151-156.

Holm, T.R. and C.D. Curtiss, 1989. A Comparison of Oxidation-Reduction Potentials Calculated from the As(V)/As(III) and Fe(III)Fe(II) Couples with Measured Platinum-Electrode Potentials in Groundwater. *Journal of Contaminant Hydrology*, 5: 67-81.

Hug, S.J., H.U. Laubscher and B.R. James, 1997. Iron(III) Catalyzed Photochemical Reduction of Chromium(VI) by Oxalate and Citrate in Aqueous Solutions. *Environmental Science & Technology*, 31(1): 160-170.

Johnson, C.A., 1990. Rapid Ion-Exchange Technique for the Separation and Preconcentration of Chromium(VI) and Chromium(III) in Fresh Waters. *Analytica Chimica Acta*, 238(2): 273-278.

Johnson, C.A. and A.G. Xyla, 1991. The Oxidation of Chromium(III) to Chromium(VI) on the Surface of Manganite (Gamma-MnOOH). *Geochimica Et Cosmochimica Acta*, 55(10): 2861-2866.

Jones, J., 2003. Groundwater Storage - the Western Experience. *Journal American Water Works Association*, 95(2): 71-+.

Kaczynski, S.E. and R.J. Kieber, 1994. Hydrophobic C-18 Bound Organic-Complexes of Chromium and Their Potential Impact on the Geochemistry of Chromium in Natural-Waters. *Environmental Science & Technology*, 28(5): 799-804.

Kalinowski, B.E. and P. Schweda, 1996. Kinetics of Muscovite, Phlogopite, and Biotite Disolution and Alteration at pH 1 - 4, Room Temperature. *Geochim. Cosmochim. Acta*, 60: 367-385.

Kent, D.B., J.A. Davis, L.C.D. Anderson and B.A. Rea, 1995. Transport of Chromium and Selenium in a Pristine Sand and Gravel Aquifer - Role of Adsorption Processes. *Water Resources Research*, 31(4): 1041-1050.

Kent, D.B., J.A. Davis, L.C.D. Anderson, B.A. Rea and T.D. Waite, 1994. Transport of Chromium and Selenium in the Suboxic Zone of a Shallow Aquifer - Influence of Redox and Adsorption Reactions. *Water Resources Research*, 30(4): 1099-1114.

Khan, F.A. and R.W. Puls, 2002. In Situ Abiotic Detoxification and Immobilization of Hexavalent Chromium. *Ground Water Monitoring and Remediation*, 23(1): 77-84.

Kimbrough, D.E., Y. Cohen, A.M. Winer, L. Creelman and C. Mabuni, 1999. A Critical Assessment of Chromium in the Environment. *Critical Reviews in Environmental Science and Technology*, 29(1): 1-46.

Klein, C. and J. Hurlbut, C. S., 1993. *Manual of Mineralogy*. John Wiley & Sons, New York, 681 pp.

Knauss, K.G. and T.J. Wolery, 1989. Muscovite Dissolution Kinetics as a Function of Ph and Time at 70 °C. *Geochim. Cosmochim. Acta*, 53: 1493-1501.

Konikow, L.F., L.L. August and C.I. Voss, 2001. Effects of Clay Dispersion on Aquifer Storage and Recovery in Coastal Aquifers. *Transport in Porous Media*, 43(1): 45-64.

Kotas, J. and Z. Stasicka, 2000. Chromium Occurrence in the Environment and Methods of Its Speciation. *Environmental Pollution*, 107(3): 263-283.

Kraemer, S., 1997. Ligand Controlled Weathering Kinetics of Aluminum Oxide and Goethite. Ph.D. Thesis, Technischen Hochschule Darmstadt, Darmstadt, 113 pp.

Kratochvil, D., P. Pimentel and B. Volesky, 1998. Removal of Trivalent and Hexavalent Chromium by Seaweed Biosorbent. *Environmental Science & Technology*, 32(18): 2693-2698.

Langmuir, D., 1997. *Aqueous Environmental Geochemistry*. Prentice Hall, Upper Saddle River, 600 pp.

Lasaga, A., 1984. Chemical Kinetics of Water-Rock Interactions. *Journal of Geophysical Research*, 89: 4009-4025.

Lasaga, A.C., 1981. Transition State Theory. In: A.C. Lasaga and R.J. Kirkpatrick (Editors), *Kinetics of Geochemical Processes*. Reviews of Mineralogy. Mineralogical Society of America, pp. 135-169.

Ledin, A., S. Karlsson, A. Duker and B. Allard, 1995. Characterization of the Submicrometer Phase in Surface Waters - a Review. *Analyst*, 120(3): 603-608.

Lee, G. and J.G. Hering, 2003. Removal of Chromium(VIi) from Drinking Water by Redox-Assisted Coagulation with Iron(II). *J. Water Supply: Res. Technol. - AQUA*, 52: 319-332.

Lee, S.Z., L.Z. Chang and R.S. Ehrlich, 1999. The Relationship between Adsorption of Cr(VI) and Soil Properties. *Journal of Environmental Science and Health Part A-Toxic/Hazardous Substances & Environmental Engineering*, 34(4): 809-833.

Lee, T., H. Lim, Y. Lee and J.W. Park, 2003. Use of Waste Iron Metal for Removal of Cr(VI) from Water. *Chemosphere*, 53(5): 479-485.

Lin, F.C. and C.V. Clemency, 1981. The Kinetics of Dissolution of Muscovites at 25°C and 1 Atm CO_2 Partial Pressure. *Geochim. Cosmochim. Acta*, 45: 571-576.

Lin, H.-C., D. Richards, C. Talbot, G.T. Yeh, J.-R. Cheng, H.-P. Cheng and N. Jones, 1997. Femwater: A Three-Dimensional Finite Element Computer Model for Simulating Density-Dependent Flow and Transport in Variably Saturated Media. Technical Report CHL-97-12, US Army Engineer Waterways Experimental Station.

Lytle, C.M., F.W. Lytle, N. Yang, J.H. Qian, D. Hansen, A. Zayed and N. Terry, 1998. Reduction of Cr(VI) to Cr(III) by Wetland Plants: Potential for in Situ Heavy Metal Detoxification. *Environmental Science & Technology*, 32(20): 3087-3093.

Ma, L. and R.F. Spalding, 1997. Effects of Artificial Recharge on Ground Water Quality and Aquifer Storage Recovery. *Journal of the American Water Resources Association*, 33(3): 561-572.

Martin, H. and A. Blankstein, 2001. Tracking Pollution May Be Tough, Los Angeles Times, Los Angeles, Jan. 31, p. B4.

Maurice, P.A., D.M. McKinght, L. Leff, J.E. Fulghum and M. Gooseff, 2002. Direct Observations of Aluminosilicate Weathering in the Hyporheic Zone of an Antarctic Dry Valley Stream. *Geochim. Cosmochim. Acta*, 66: 1335-1347.

McCarthy, J.F. and J.M. Zachara, 1989. Subsurface Transport of Contaminants - Mobile Colloids in the Subsurface Environment May Alter the Transport of Contaminants. *Environmental Science & Technology*, 23(5): 496-502.

McQuarrie, J.P. and K. Carlson, 2003. Secondary Benefits of Aquifer Storage and Recovery: Disinfection by-Product Control. *Journal of Environmental Engineering-ASCE*, 129(5): 412-418.

Mirecki, J.E., B.G. Campbell, K.J. Conlon and M.D. Petkewich, 1998. Solute Changes During Aquifer Storage Recovery Testing in a Limestone Clastic Aquifer. *Ground Water*, 36(3): 394-403.

Murdoch, J. and R.W. Webb, 1966. *Minerals of California*, 189. Calif. Division of Mines and Geology, San Francisco, 559 pp.

Murphy, D.M., J.R. Garbarino, H.E. Taylor, B.T. Hart and R. Beckett, 1993. Determination of Size and Element Composition Distributions of Complex Colloids by Sedimentation Field-Flow Fractionation Inductively-Coupled Plasma-Mass Spectrometry. *Journal of Chromatography*, 642(1-2): 459-467.

Nickel, E., 1973. Experimental Dissolution of Light and Heavy Minerals in Comparison with Weathering and Intrastratal Solution. *Contrib. Sediment.*, 1: 1-68.

NTP, 2003. Planned Studies on Hexavalent Chromium. National Toxicity Program Fact Sheet, http://ntp-server.niehs.nih.gov/.

Oelkers, E.H. and J. Schott, 2001. An Experimental Study of Enstatite Dissolution Rates as a Function of pH, Temperature, and Aqueous Mg and Si Concentration, and the Mechanism of Pyroxene/Pyroxenoid Dissolution. *Geochimica et Cosmochimica Acta*, 65: 1219-1231.

Paces, T., 1983. Rate Constants of Dissolution Derived from Measurements of Mass Balance in Hydrological Catchments. *Geochimica et Cosmochimica Acta*, 47: 1855-1863.

Parkhurst, D.L. and M.D. Petkewich, 2002. Geochemical Modeling of an Aquifer Storage Recovery Experiment, Charleston, South Carolina. In: G.R. Aiken and E.L. Kuniansky (Editors), U.S. Geological Survey Artifical Recharge Workshop Proceedings, Sacremento, CA, April 2-4. U.S. Geological Survey.

Paustenbach, D.J., B.L. Finley, F.S. Mowat and B.D. Kerger, 2003. Human Health Risk and Exposure Assessment of Chromium (VI) in Tap Water. *Journal of Toxicology and Environmental Health-Part A*, 66(14): 1295-1339.

Pavel, J., J. Kliment, S. Stoerk and O. Suter, 1985. Preservation of Traces of Chromium (VI) in Water and Waste- Water Samples. *Fresenius Zeitschrift Fur Analytische Chemie*, 321(6): 587-591.

Pettine, M., L. D'Ottone, L. Campanella, F.J. Millero and R. Passino, 1998. The Reduction of Chromium (VI) by Iron (II) in Aqueous Solutions. *Geochimica Et Cosmochimica Acta*, 62(9): 1509-1519.

Philipot, J.M., F. Chaffange and J. Sibony, 1985. Hexavalent Chromium Removal from Drinking-Water. *Water Science and Technology*, 17(6-7): 1121-1132.

Powell, R.M., R.W. Puls, S.K. Hightower and D.A. Sabatini, 1995. Coupled Iron Corrosion and Chromate Reduction - Mechanisms for Subsurface Remediation. *Environmental Science & Technology*, 29(8): 1913-1922.

Puls, R.W., C.J. Paul and R.M. Powell, 1999. The Application of in Situ Permeable Reactive (Zero-Valent Iron) Barrier Technology for the Remediation of Chromate- Contaminated Groundwater: A Field Test. *Applied Geochemistry*, 14(8): 989-1000.

Pyne, R.D.G., 1995. *Groundwater Recharge and Wells*. Lewis, Boca Raton, FL, 376 pp.

Rai, D., L.E. Eary and J.M. Zachara, 1989. Environmental Chemistry of Chromium. *Science of the Total Environment*, 86(1-2): 15-23.

Rai, D., B.M. Sass and D.A. Moore, 1987. Chromium(III) Hydrolysis Constants and Solubility of Chromium(III) Hydroxide. *Inorganic Chemistry*, 26(3): 345-349.

Reddad, Z., C. Gerente, Y. Andres and P. Le Cloirec, 2003. Mechanisms of Cr(III) and Cr(VI) Removal from Aqueous Solutions by Sugar Beet Pulp. *Environmental Technology*, 24(2): 257-264.

Reese, R.S., 2002. Inventory and Review of Aquifer Storage and Recovery in Southern Florida. WRI 02-4036, U.S. Geological Survey, Tallahassee, FL.

Reinhard, M., J. Montgomery-Brown, J.S. Louie and B. Gross, 2003. From Effluent to New Water: Performance Evaluation and Quality Assurance. *Chimia*, 57(9): 561-566.

Rengaraj, S., C.K. Joo, Y. Kim and J. Yi, 2003. Kinetics of Removal of Chromium from Water and Electronic Process Wastewater by Ion Exchange Resins: 1200h, 1500h and Irn97h. *Journal of Hazardous Materials*, 102(2-3): 257-275.

Richard, F.C. and A.C.M. Bourg, 1991. Aqueous Geochemistry of Chromium - a Review. *Water Research*, 25(7): 807-816.

Robertson, F.N., 1975. Hexavalent Chromium in the Ground Water in Paradise Valley, Arizona. *Ground Water*, 13(6): 516-527.

Robles-Camacho, J. and M.A. Armienta, 2000. Natural Chromium Contamination of Groundwater at Leon Valley, Mexico. *Journal of Geochemical Exploration*, 68(3): 167-181.

Ruan, F. and D. McLaughlin, 1999. An Investigation of Eulerian-Lagrangian Methods for Solving Heterogenous Advection-Dominated Transport Problems. *Water Resources Research*, 35: 2359-2373.

Schecher, W.D. and D.C. McAvoy, 1998. Mineql+: A Chemical Equilibrium Modeling System, Environmental Research Software, Hallowell.

Schlautman, M.A. and I. Han, 2001. Effects of pH and Dissolved Oxygen on the Reduction of Hexavalent Chromium by Dissolved Ferrous Iron in Poorly Buffered Aqueous Systems. *Water Research*, 35(6): 1534-1546.

Schott, J. and R.A. Berner, 1983. X-Ray Photoelectron Studies of the Mechanisms of Iron Silicate Dissolution During Weathering. *Geochim. Cosmochim Acta*, 47: 2233-2240.

Schroeder, D.C. and G.F. Lee, 1975. Potential Transformations of Chromium in Natural Waters. *Water, Air, Soil Poll.*, 4: 355-365.

Sedlak, D.L. and P.G. Chan, 1997. Reduction of Hexavalent Chromium by Ferrous Iron. *Geochimica Et Cosmochimica Acta*, 61(11): 2185-2192.

Sorg, T.J., 1979. Treatment Technologies to Meet the Interim Primary Drinking Water Regulations for [in]Organics: Part 4. *J. Amer. Water Works Assoc.*, 71(8): 454-466.

Sreeram, K.J. and T. Ramasami, 2001. Speciation and Recovery of Chromium from Chromite Ore Processing Residues. *Journal of Environmental Monitoring*, 3: 526-530.

Stollenwerk, K.G. and D.B. Grove, 1985. Adsorption and Desorption of Hexavalent Chromium in an Alluvial Aquifer near Telluride, Colorado. *Journal of Environmental Quality*, 14(1): 150-155.

Stumm, W., 1992. *Chemistry of the Solid-Water Interface*. Wiley-Interscience, New York, 428 pp.

Stumm, W. and R. Wollast, 1990. Coordination Chemistry of Weathering: Kinetics of the Surface-Controlled Dissolution of Oxide Minerals. *Rev. Geophys.*, 28: 53-69.

Swoboda-Colberg, N.G. and J.I. Drever, 1993. Mineral Dissolution Rates in Plot-Scale Field and Laboratory Experiments. *Chemical Geology*, 105: 51-69.

Tahir, H., M. Saleem, M. Afzal, H. Ahmad, S.T. Hussain and J. Afzal, 1998. Estimation and Removal of Chromium Ions from Tannery Wastes Using Zeolite-3a. *Adsorption Science & Technology*, 16(3): 153-161.

Taylor, H.E., J.R. Garbarino, D.M. Murphy and R. Beckett, 1992. Inductively Coupled Plasma Mass-Spectrometry as an Element- Specific Detector for Field-Flow Fractionation Particle Separation. *Analytical Chemistry*, 64(18): 2036-2041.

Tokunaga, T.K., J.M. Wan, M.K. Firestone, T.C. Hazen, K.R. Olson, D.J. Herman, S.R. Sutton and A. Lanzirotti, 2003. In Situ Reduction of Chromium(VI) in Heavily Contaminated Soils through Organic Carbon Amendment. *Journal of Environmental Quality*, 32(5): 1641-1649.

Tzou, Y.M., M.K. Wang and R.H. Loeppert, 2003. Sorption of Phosphate and Cr(VI) by Fe(III) and Cr(III) Hydroxides. *Arch. Environ. Contam. Toxicol.*, 44: 445-453.

Vainshtein, M., P. Kuschk, J. Mattusch, A. Vatsourina and A. Wiessner, 2003. Model Experiments on the Microbial Removal of Chromium from Contaminated Groundwater. *Water Research*, 37(6): 1401-1405.

Van Geen, A., A.P. Robertson and J.O. Leckie, 1994. Complexation of Carbonate Species at the Goethite Surface: Implications for Adsorption of Metal Ions in Natural Waters. *Geochim. Cosmochim. Acta*, 58: 2073-2086.

van Genuchten, M.T.A., 1980. Closed-Form Equation for Predicting the Hydraulic Conductivity of Unsaturated Soils. *Soil Science Society of America J.*, 44: 892-898.

Velbel, M.A., 1993. Constancy of Silicate-Mineral Weathering-Rate Ratios between Natural and Experimental Weathering: Implications for Hydrologic Control of Differences in Absolute Rates. *Chemical Geology*, 105: 89-99.

Villalobos, M., M.A. Trotz and J.O. Leckie, 2001. Surface Complexation Modeling of Carbonate Effects on the Adsorption of Cr(VI), Pb(II), and U(VI) on Goethite. *Environ. Sci. Technol.*, 35: 3849-3856.

Vincent, J.B., 2000. The Biochemistry of Chromium. *Journal of Nutrition*, 130(4): 715-718.

Weaver, R.M., M.F. Hochella and E.S. Ilton, 2002. Dynamic Processes Occurring at the Cr-aq(III)-Manganite (Gamma- MnOOH) Interface: Simultaneous Adsorption, Microprecipitation, Oxidation/Reduction, and Dissolution. *Geochimica Et Cosmochimica Acta*, 66(23): 4119-4132.

Wen, L.S., M.C. Stordal, D.G. Tang, G.A. Gill and P.H. Santschi, 1996. An Ultraclean Cross-Flow Ultrafiltration Technique for the Study of Trace Metal Phase Speciation in Seawater. *Marine Chemistry*, 55(1-2): 129-152.

White, A.F., A.E. Blum, M.S. Schulz, T.D. Bullen, J.W. Harden and M.L. Peterson, 1996. Chemical Weathering Rates of a Soil Chronosequence on Granitic Alluvium. 1. Quantification of Mineralogical and Surface Area Changes and Calculation of Primary Silicate Reaction Rates. *Geochimica et Cosmochimica Acta*, 60(14): 2533-2550.

White, A.F. and A. Yee, 1985. Aqueous Oxidation-Reduction Kinetics Associated with Coupled Electron-Cation Transfer from Iron-Containing Silicates at 25°C. *Geochim. Cosmochim. Acta*, 49: 1263-1275.

Wieland, E. and W. Stumm, 1992. Dissolution Kinetics of Kaolinite in Acidic Aqueous Solutions at 25 °C. *Geochim. Cosmochim. Acta*, 56: 3339-3355.

Wittbrodt, P.R. and C.D. Palmer, 1996. Effect of Temperature, Ionic Strength, Background Electrolytes, and Fe(Iii) on the Reduction of Hexavalent Chromium by Soil Humic Substances. *Environmental Science & Technology*, 30(8): 2470-2477.

Yalcin, S., R. Apak, J. Hizal and H. Afsar, 2001. Recovery of Copper (Ii) and Chromium (Iii,Vi) from Electroplating-Industry Wastewater by Ion Exchange. *Separation Science and Technology*, 36(10): 2181-2196.

Yeh, G.-T., K.M. Salvage, J.P. Gwo, J.M. Zachara and J.E. Szecsody, 1998. Hydrobiogeochem: A Coupled Model of Hydrologic Transport and Mixed Biogeochemical Kinetic/Equilibrium Reactions in Saturated-Unsaturated Media. Report ORNL/TM-13668, Oak Ridge National Laboratory,, Oak Ridge, TN.

Zachara, J.M., D.C. Girvin, R.L. Schmidt and C.T. Resch, 1987. Chromate Adsorption on Amorphous Iron Oxyhydroxide in the Presence of Major Groundwater Ions. *Environmental Science & Technology*, 21(6): 589-594.

Zhang, M. and E.J. Reardon, 2003. Removal of B, Cr, Mo, and Se from Wastewater by Incorporation into Hydrocalumite and Ettringite. *Environmental Science & Technology*, 37(13): 2947-2952.

Zotter, K. and I. Licsko, 1992. Removal of Chromium(VI) and Other Heavy Metals from Groundwaters in Neutral and Alkaline Media. *Wat. Sci. Tech.*, 26(1-2): 207-216.

ABBREVIATIONS

A	specific surface area of solid
A_e	elements area
A_{elem}	area of recharge boundary element face
A_m	surface area of mineral *m*
a_L	longitudinal dispersivity
a_M	molecular diffusion coefficient
a_T	lateral dispersivity
AAS	atomic absorption spectroscopy
addl	additional
ASR	aquifer storage and recovery
AwwaRF	Awwa Research Foundation
α	soil-specific coefficient
α'	compressibility of the medium
BET	Brunauer Emmet Teller
β	soil-specific exponent
C_i	concentration of species i
C_{in}	material concentration in the source
$\hat{C}(x,y,z,t)$	concentration value (above background) specified in space and time
cit.	cited
cm	centimeter
Co.	company
conc	concentration
Cr_T	total chromium concentration
CRW	Colorado river water
d	day
Dept.	department
Dep Vars	dependent variables
dm^3	cubic decimeter (1 liter)
DOC	dissolved organic carbon
°C	degree Celsius
° 2Θ	degree 2 theta
∇	del operator
δ	Kronecker delta tensor
Δt	time step
EDTA	ethylenediaminetetraacetate
EPA	Environmental Protection Agency
F	storage coefficient
Fe_T	total iron concentration

g	gram
g	acceleration of gravity
GB	gigabytes
GHz	gigahertz
g/L	gram per liter
g L^{-1}	gram per liter
GMS	Department of Defense Groundwater Modeling System
γ	soil-specific exponent
h	hour
h	pressure head
HAc	acetic acid
HAH	hydroxylamine hydrochloride
HBGC123D	HydroBioGeoChem model for 1, 2, or 3 Dimensions
HDPE	high density polyethylene
i_{max}	maximum level of resolution
ICP	inductively coupled plasma
ICP-MS	inductively coupled plasma mass spectrometry
ID	identification
Inc.	incorporated
k	rate constant
k_{ads}	adsorption rate constant
K_D	distribution coefficient
k_{des}	desorption rate constant
k_{im}	rate constant for dissolution of species *i* in mineral *m*
k_m	rate constant for dissolution of mineral *m*
k_{ox}	oxidative dissolution rate constant
k_r	relative permeability
$\mathbf{k_s}$	saturated permeability tensor
K	coefficient in Langmuir or Freundlich isotherm
K_d	distribution coefficient
K_s	first-order biogradation rate constant in sorbed phase
K_w	first-order biogradation rate constant in dissolved phase
kg	kilogram
kJ	kilojoules
KV	kilovolt
L	liter
L $\mu g^{-1} min^{-1}$	liter per microgram per minute
λ	decay constant
m	artifical mass rate
m	meter

$\tilde{m}_c$	constituent mass
m^2	square meter
m^3	cubic meter
M	molar
M_{ijk}	ijk^{th} moment of the concentration distribution in space
M_{low}	lower mass limit
$[M]_{ads}$	adsorbed metal concentration
$[M]_{diss}$	dissolved metal concentration
$[M]_T$	total dissolved metal concentration
mA	milliampere
MCL	maximum contaminant level
mg/L	milligram per liter
m^2/g	square meter per gram
$m^2\ g^{-1}$	square meter per gram
Milli-Q water	water purified using a Millipore Milli-Q 18 MΩ system
min	minute
min^{-1}	per minute
mL	milliliter
$mL\ h^{-1}$	milliliter per hour
mL/h	milliliter per hour
mL/min	milliliter per minute
mM	millimolar
mm	millimeter
mol	mole
$mol\ L^{-1}$	mol per liter
$mol\ L^{-1}time^{-1}$	mol per liter per time
$mol\ m^{-2}s^{-1}$	mol per square meter per second
MWDSC	Metropolitan Water District of Ssouthern California
μ	dynamic viscosity of water
μM	micromolar
μm	micrometer
$\mu g\ L^{-1}$	microgram per liter
$\mu g\ L^{-1}min^{-1}$	microgram per liter per minute
μg/g	microgram per gram
μg/L	microgram per liter
μmol/g	micromole per gram
μmol/L	micromole per liter
n	power index in the Freundlich isotherm
N	normality (equivalents per liter)
v_i	stoichiometric content of species *i* in mineral *m*
P	percent difference
PAC	project advisory committee
%	per cent

q	flow rate
q	source and/or sink
Q_r	recharge rate
R	steady state release rate
r	ratio
ref.	reference
rpm	revolution per minute
ρ	density of water
ρ_b	bulk density of the medium
S	solid concentration
S	material concentration in the adsorbed phase
S_{max}	maximum concentration of the medium in a Langmuir isotherm
$[S]$	free surface site concentration
s	second
site/nm^2	site pre square nanometer
$[S]_T$	total surface site concentration
t	time
$time^{-1}$	per time
TOL	tolerance level
τ	tortuosity
θ	moisture concentration
θ_w	moisture content
θ_e	effective moisture content
θ_r	residual moisture content
θ_s	saturation moisture content
USGS	United States Geological Survey
V	reactor volume
V	discharge
V	volume of solution
V	Darcy velocity
v/v	volume per volume
wt %	weight per cent
x_c	x-coordinate location of the center of mass of the recharged water plume
X_{diss}	dissolved X
X_{ads}	adsorbed X
$X_{1\text{-}solid}$	X in primary solid
$X_{2\text{-}solid}$	X in secondary solid

XRD	x-ray diffraction
XRF	x-ray fluorescence
y_c	y-coordinate location of the center of mass of the recharged water plume
z	potential head
z_c	z-coordinate location of the center of mass of the recharged water plume

[illegible] EU representative [illegible]
[illegible]
[illegible]

Printed and bound by CPI Group (UK) Ltd, Croydon, CR0 4YY
[illegible]

IWA Publishing's authorised EU representative for General Product Safety Regulations is Diane D'Arras, 15 rue Duret, 75116 Paris, France, e-mail: safety@iwap.co.uk.

Printed and bound by CPI Group (UK) Ltd, Croydon, CR0 4YY
17/09/2026
02218811-0016